Birthstone
of the White House and Capitol

Jane Hollenbeck Conner

Best Wishes to Tom Kenavan, a longtime Aquia Harbour resident.

Jane Conner

Copyright (c) 2005 by Jane Hollenbeck Conner

All rights reserved, including the right to reproduce this work in any form whatsoever without permission in writing from the publisher, except for brief passages in connection with a review.

For information, please write:
The Donning Company Publishers
184 Business Park Drive, Suite 206
Virginia Beach, Virginia 23462–6533

Steve Mull, General Manager
Barbara Buchanan, Office Manager
Kathleen Sheridan, Senior Editor
Lynn Parrott, Graphic Designer
Amy Thomann, Imaging Artist
Mary Ellen Wheeler, Proofreader
Scott Rule, Director of Marketing
Stephanie Linneman, Marketing Coordinator
Lori Porter, Project Research Coordinator

Dennis Walton, Project Director

Library of Congress Cataloging-in-Publication Data
Conner, Jane Hollenbeck.
  Birthstone of the White House and Capitol / Jane Hollenbeck Conner.
    p. cm.
  Includes bibliographical references and index.
  ISBN 1-57864-336-8 (alk. paper)
 1. United States Capitol (Washington, D.C.)--History. 2. White House (Washington, D.C.)--
History. 3. Sandstone buildings--Washington (D.C.)--History. 4. Sandstone buildings--
Conservation and restoration--Washington (D.C.)--History. 5. Sandstone--Virginia--Aquia
Creek--History. 6. Architecture--Washington (D.C.)--18th century. 7. Architecture--Washington
(D.C.)--19th century. 8. City planning--Washington (D.C.)--History. 9. Architecture and state--
United States--History. 10. Washington (D.C.)--Buildings, structures, etc. I. Title.
  F204.C2C66 2005
  975.3'02--dc22

                              2005028364

Printed in the United States of America by Walsworth Publishing Company

Front cover photos: White House and Capitol by
G. Wang, University of Central Florida
Back cover photos: Columns © 1992 Southern Living, Inc.
Reprinted with permission. Carving over White House by
Jack Boucher, Historic American Buildings Survey
Background photo: Barbara S. Kirby

Dedicated to

Rex Scouten

*who served under ten U.S. Presidents,
for his steadfast and earnest support of this work and
other efforts to preserve Government Island*

Birthstone
of the White House and Capitol

Jane Hollenbeck Conner

Contents

	Acknowledgments	6
	Prologue	7
chapter 1	What Is Aquia Freestone?	8
chapter 2	Aquia Stone Boundary Markers for the New Capital	16
chapter 3	Who Chose the Stone?	22
chapter 4	L'Enfant and the City's Genesis	28
chapter 5	Jefferson, Brick vs. Stone	32
chapter 6	Cornerstone Ceremonies	38
chapter 7	Many Workers Are Needed: Slave, Free, and Skilled	44
chapter 8	Quarrying Aquia Stone	52
chapter 9	Transporting the Stone	58
chapter 10	Work Continues	66
chapter 11	Moving into the Federal Buildings	74
chapter 12	Benjamin Henry Latrobe's Contributions	78
chapter 13	Other Freestone Quarries	84
chapter 14	Cutting and Carving Aquia Stone	92
chapter 15	War of 1812	104

chapter 16 — 110
Rebuilding

chapter 17 — 122
Capitol's Completion and Quarries' Closure

chapter 18 — 134
Truman's White House Renovation

chapter 19 — 142
Extending the Capitol's East Front

chapter 20 — 146
Who Owned Government Island?

chapter 21 — 150
Deterioration of Aquia Stone

chapter 22 — 154
Painting and Restoring the Stone of the White House

chapter 23 — 164
Removing and Replacing the Capitol's Stone

Epilogue — 168

Appendix I   Rock of Ages — 169

Appendix II   Other Structures of Aquia Stone — 175

Notes — 192

Bibliography — 214

Index — 219

Acknowledgments

Rex W. Scouten, former White House Curator. Dr. Robert J. Kapsch, National Park Service, Senior Scholar in Historic Engineering and Architecture (RET.) Patrick J. Plunkett, Master Stone Carver and Mason. William C. Allen, Architectural Historian. Dr. Barbara A. Wolanin, Curator, and Ann Kenny, Administrative Assistant, Office of the Architect of the Capitol. Jack E. Boucher, HABS. Bill Allman, Curator of the White House. Gary J. Walters, White House Chief Usher. C. M. Williams, former Stafford County Administrator. Barbara Schomp Kirby and Stephen Gambaro, Stafford, VA. John Riley, White House Historical Association. Bob Arnebeck, Washington, D.C. Kenneth R. Bowling, First Federal Congress Project, George Washington University. Noel Harrison, Charlottesville, VA. John and Jan Zweifel, The White House Exhibition, Orlando, FL. C. Michael Flanagan. George Gordon, Thomas Metts, Charles Price, Wilbur Segar, and the late Milton Dickerson, Stafford, VA. Carson Rhyne, Richmond. Susan Bouchard, Gunston Hall. Matt Webster, George Washington's Fredericksburg Foundation. John Pearce, the James Monroe Museum. Jim Hall and Lou Cordero, Fredericksburg, VA. Pamela Scott, L'Enfant Papers. Steve Crosby, Stafford County Administrator. Mary Ruth Coleman, The Carlyle House. Stafford County Board of Supervisors. David Noel, Stafford. Douglas A. Brown, CA. My dear mother, Margie Mild Hollenbeck, who instilled in me a love of history. My wonderful children, Michelle Porter and Douglas Henderson, who put up with my frequent trips to the island. My dear husband, Al Conner, who spent hours proofreading, scanning, and encouraging.

Prologue

The White House and the United States Capitol are the two most widely recognizable buildings in the United States. They grace our stamps, coins, and paper currency and are recognized throughout the world as symbols of freedom and liberty.

These two buildings instill pride in countless Americans. The White House, although it contains executive offices and the President's private residence, is considered "The House of the People." First Lady Abigail Adams spoke of its timeless quality when she penned, "…this House is built for ages to come."[1] Nathaniel Hawthorne, visiting Washington in 1862, immediately grasped the significance of the Capitol to a war-torn nation when he wrote:

> *It is natural enough to suppose that the center and heart of America is the Capitol and certainly, in its outward aspect, the world has not many statelier or more beautiful edifices….*

Noted historian Allan Nevins wrote that the Capitol is "the spirit of America in stone."[2]

The stone that was used to create these two landmarks was quarried in Virginia. This is an account of its origin and its unique place in American history, spanning over 200 years and combining the actions of great political figures, leading architects, immigrant artisans, and free and enslaved laborers. The stone's story expands into the twenty-first century, chronicling efforts to preserve its history for future generations.

1

What Is Aquia Freestone?

A scholarly, middle-aged gentleman was touring the United States Capitol. After examining the interior walls, he raised his hand and asked, "Are these walls made of freestone?" The young tour guide responded by saying, "No, sir. The walls are Aquia sandstone." Actually, the gentleman was correct; freestone was the term used more than 200 years ago for the sandstone that created both the Capitol and the White House. George Washington, Thomas Jefferson, Pierre Charles L'Enfant, and others responsible for the construction of the nation's capital buildings never used the term sandstone but instead wrote "free stone" or "freestone." Freestone was a term given to a stone that could be cut or carved freely without splitting.

Aquia freestone, or sandstone, was found along a creek of the same name in Stafford County, Virginia. Aquia (pronounced ah qui' ah) Creek is located about forty miles south of Washington, D.C., and flows into the Potomac River. The original name was "Quiyough" and was given to the creek by Captain John Smith when he visited the area in 1608. Smith, in his writings, used the name Quiyough twice—once for the Indian village located at the south side of the creek and once for the creek itself.[1] According to Indian language authorities,

Birthstone of the White House and Capitol

the word is Algonquin and means "gulls." Maps later corrupted the word to "Quia" and still later to "Acquia."

Aquia freestone is an arkose sandstone of the Lower Cretaceous Period. This simply means it was formed 100 million years ago to 136 million years ago. Arkose identifies it as a sandstone containing 25 percent or more feldspar.[2]

This aerial view shows Government Island in the foreground with the residential community of Aquia Harbour in the background.
Lou Cordero

Sandstone, a sedimentary rock, was created when rocks such as granite or gneiss disintegrated, creating small pieces of quartz sand. The pieces were transported by wind or streams and deposited into river bottoms, lakes, or oceans. Pressure was exerted on these layers over millions of years. The quartz sand became united with a cement such as silica, calcium carbonate, or iron oxide. The color of the sandstone was determined by the cementing material. Iron oxide, for example, produced a red or reddish-brown stone. Other materials produced a white, yellow, or gray sandstone.

Aquia sandstone was unique for its light color, strength, and fine-grained appearance. Unlike all other sandstones, the matrix, or the material in which each grain was embedded, was harder than its crystals. Thus, it was a rock that had the quality of being hard, yet not brittle.[3]

When sandstone is carved, cut, or broken, the cement is fractured, and each little grain of sand remains whole, thus giving the surface a granular appearance. Most Aquia stone was special since it was very fine-grained. Some sandstone shipped from private quarries for the nation's buildings was rejected because it did not meet the demanding standards of the founding fathers for the new federal city. Some was too granular, some contained clay holes, some was the wrong color, and some was too soft.

GOVERNMENT ISLAND

Outcroppings of freestone or Aquia stone were and still can be found along Aquia Creek and the Rappahannock River. But most of the stone used in the district was from a small island and a few adjoining quarries located close to the creek. The island, known today as Government Island, is located about six miles from the juncture of Aquia Creek and the Potomac River.

On the island, uniformly cut quarry cliffs rise twenty to fifty feet in the air. Their surfaces expose the chisel marks created more than two centuries ago.
Barbara Kirby

Government Island is nestled close to the shore near the present-day residential community of Aquia Harbour. It appears to be an extension of land, but actually a marshy bog joins its western shore to the mainland. Boaters leaving the Aquia Harbour marina cannot see the stone; there is sufficient soil around the quarries to support the growth of trees and other flora. Even aerial photographs of the island cannot capture the abandoned stone quarries that dot its surface.

THE BRENT FAMILY

The first time history recorded the island was in a November 28, 1678, land patent. Two gentlemen received the patent from Herbert Jeffreys, the "Govenour and Capt General of Virginia," for his Royal Majesty King Charles II of England. The island was identified as "twelve Acres of Land *(today the island is recorded as being 17.389 acres)* lyeing in Md County with a small point of Marsh joynineg to it lying in Acquia Creek...."[4]

Evidently, the two men did not follow the agreement of the land patent; on January 30, 1694, a deed for the island mentions that the 1678 patent was void. This deed stated that a George Brent "hast purchased a small tongue neck or Island of Land with small point of marsh...."[5]

George Brent and his descendants held the island for almost one hundred years. Their control of the island quarries helped shape the use of the stone in buildings and structures throughout colonial America and eventually in the new independent nation.

George Brent was a member of the colorful Brent family that played an important role in the early history of Virginia. His uncle, Colonel Giles Brent of Maryland, was a gentleman of Catholic faith who had grown weary of the Puritans' bickering in that colony. Since he was also dissatisfied with the Calvert regime, he crossed the Potomac River in 1647 with his Piscataway Indian princess wife and settled on a northern peninsula of Virginia at the mouth of Aquia Creek. He was followed to the Aquia area by his sisters, Mary and Margaret.

Margaret Brent was the first female attorney in the New World. She was also considered by many as the first suffragette in America because she demanded a vote in the Maryland Royal Council on the grounds that she was the sole executrix of the late Honorable Leonard Calvert, Governor of Maryland. She acted as attorney for her brother and sister and became the first woman in America to hold land in her own right. She owned land not only in Maryland but also in Virginia. She held the land that is now Fredericksburg and Alexandria, Virginia.[6]

George Brent joined this interesting family in 1673 when he was sent from England to reside with his uncle "to learn how to live." He certainly accomplished this; he later became captain of the Militia, lawyer, attorney general of the Colony, and representative to the Virginia legislature. George established Woodstock Plantation on upper Aquia Creek, the site of the present-day community of Aquia Harbour.[7]

WIGGINGTON'S ISLAND

In the 1694 document that gave George Brent the island, the name Wiggington Island was added to the deed in red ink. It is not known where this name originated. There are no Wiggingtons mentioned in any chain of title. A William Wiggington was mentioned in the Register of Overwharton Parish. The parish church, located close to the island, collected the taxes for the crown, and the records show the amount of tobacco that was collected. According to the register, William Wiggington was an early settler of Aquia Creek whose wife and three children were murdered by a band of Indians from Maryland in 1697.[8] Records, however, do not indicate if he or his ancestors ever lived on or owned the island. The name Wiggington's Island, nonetheless, was used on some early documents and stayed with the island for decades.

Historical documentation for the island is scarce. One can only speculate as to its appearance in the late 1600s and 1700s and to its use as a stone quarry. Existing Brent family papers do not mention the island's early appearance, and any official documents concerning its use as a quarry were probably burned in the 1730 fire of the Stafford County Courthouse. Judging from the remaining tall cliffs of the abandoned quarry sites and considering the vast amounts of stone quarried for the Washington, D.C., buildings, the buff-colored sandstone island rising from Aquia Creek must have been an impressive sight.

Another early use of the quarried stone was that of making tombstones. Aquia stone grave markers found at the Brent family cemetery date back to 1681.

FIRST USES OF FREESTONE

With freestone so plentiful, it was probably first quarried for building material. The Brent family, although Roman Catholic, was broadly tolerant and allowed people of all religions, including Cavaliers and later Huguenot refugees, to settle in their Aquia Creek villages in the late 1600s. (This was the first instance of religious toleration in Virginia.) The influx of people created a need for housing. Brent's own Woodstock Plantation home had a freestone basement, so presumably other homes in the area had freestone foundations, too.

In the early and middle 1700s, most Virginia houses were made of brick and wood rather than stone. Plantation owners tended to make their houses along simple lines, but if cut stone was employed, it was usually done sparingly. If such was the case, the stone was either imported from England, or Aquia stone from the Brent quarry was used.[9]

Aquia Church, near Wiggington's Island, was built in 1751. It is an excellent standing example of how freestone and brick were combined. The church, built in cruciform design has eight corners each trimmed with Aquia stone quoins. Keystones of freestone were inserted in the round brick arches of the windows. The doorways were decorated with sandstone instead of rubbed brick, which was so popular in the Virginia Tidewater

Birthstone of the White House and Capitol

Many homes and churches, such as Aquia Church, built in 1751, used Aquia stone for architectural trim.
Historic American Buildings Survey/Library of Congress

churches. Other churches in Virginia used the stone in combination with brick. George Washington's Pohick Church in Lorton, Virginia, as well as his Christ Church in Alexandria, Virginia, used Aquia stone as trim.

Chimney stones or pieces were also made of freestone during the eighteenth century. Contrary to its name, a chimney stone was not placed on the exterior of the house but was actually placed in the interior, for it was a decorative mantel piece. George Mason, author of the Virginia Declaration of Rights on which America's Bill of Rights was based, wrote to his son about obtaining chimney stones or pieces from Aquia.[10]

The Brent's stone quarry not only did business with Virginians but also shipped stone to Philadelphia to make buildings and bridges. A letter from Daniel Carroll Brent to his brother, Robert, states, "A load of stone would have been sent some time ago to Philadelphia, could a vessel have been got; when one is to be got, it will be shipped."[11]

For more than a century, freestone from Aquia not only was used to make gravestones, build and trim homes and churches, and construct bridges, but also was used for steppingstones, millstones, and steps. Even George Washington used the Brent quarried stone for his Mount Vernon steps. But the most significant use of all came in 1791, when it became the major building block of Washington, D.C.

What Is Aquia Freestone?

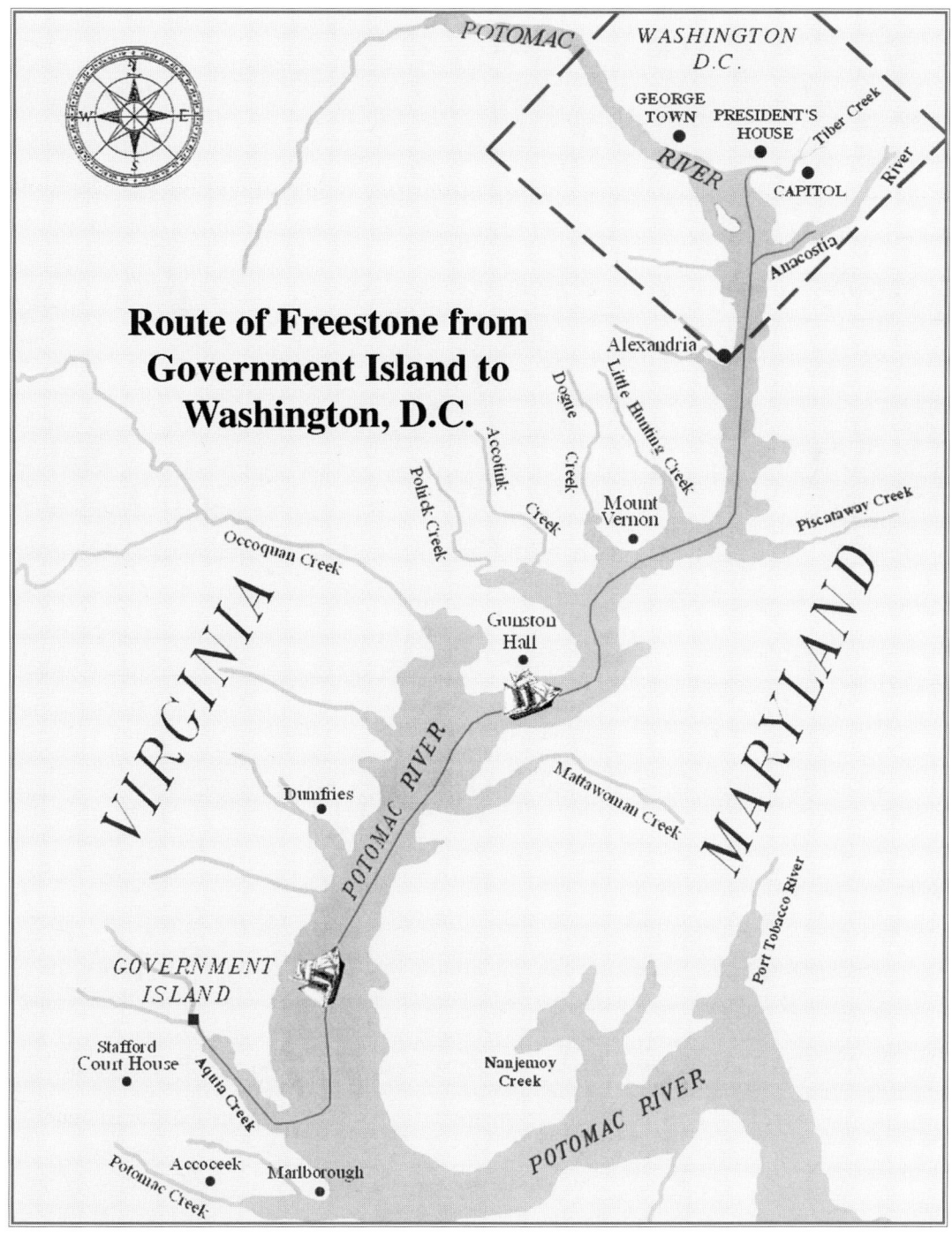

Chuck Nelson, GeoMetrics GPS Inc., Stafford, Virginia

2

Aquia Stone Boundary Markers for the New Capital

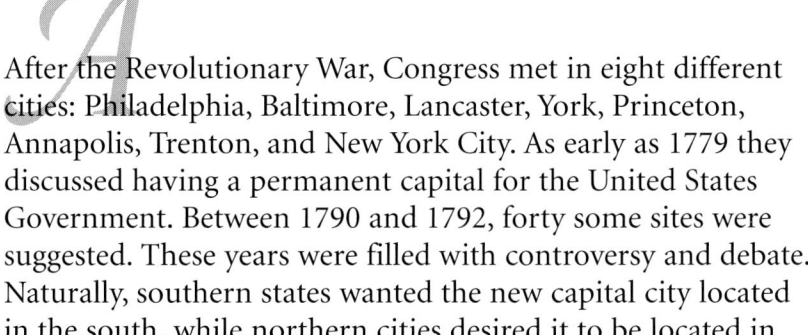

After the Revolutionary War, Congress met in eight different cities: Philadelphia, Baltimore, Lancaster, York, Princeton, Annapolis, Trenton, and New York City. As early as 1779 they discussed having a permanent capital for the United States Government. Between 1790 and 1792, forty some sites were suggested. These years were filled with controversy and debate. Naturally, southern states wanted the new capital city located in the south, while northern cities desired it to be located in the north.

In 1790, President Washington asked Congress to let him choose a site on the Potomac "not exceeding 10 miles square" for the new capital. The site he chose included part of Maryland, a northern state, and part of Virginia, a southern state. (The exact location chosen by Washington had been proposed two years before by a Scottish merchant of Georgetown by the name of George Walker.[1]) Being in both states, this compromise site appeased the debating factions. The area also included the two existing cities of Alexandria and Georgetown and afforded needed transportation since it was lying on the Potomac River.

A year later Congress authorized George Washington to appoint three Commissioners. These gentlemen would survey the site and oversee the construction and design of the new city and federal buildings. Washington appointed three intimate friends:

Birthstone of the White House and Capitol

- Thomas Johnson, like Washington, served as general in the Revolutionary War. He was a lawyer from Maryland who became the first governor of the state. Both Johnson and Washington were fifty-nine.

- Daniel Carroll, sixty-one, was also from Maryland. Brother of the first American Roman Catholic bishop, Carroll was extremely well educated and had been a Representative in Congress. He owned thousands of acres of land along the Potomac and had two lots in the new federal city.

- David Stuart was from Virginia. He knew Washington well, as he married Martha Washington's widowed daughter-in-law. Stuart, in his late thirties, was a practicing physician in Alexandria, Virginia, as well as a planter who had served in the Virginia legislature.[2]

A vote on the final boundaries also occurred in 1791. At that time Andrew Ellicott, a noted surveyor, was chosen to plot the diamond-shaped area. Forty Aquia sandstone markers would be placed one mile apart, creating the boundary for the new nation's capital.

On April 15, 1791, Ellicott, the capital Commissioners, and the mayor of Alexandria, Virginia, along with various local dignitaries, went to Jones Point, located near today's Woodrow Wilson Bridge. Freemasons, in an elaborate ceremony, placed the first boundary stone on the southernmost corner. Its exact location was calculated by Ellicott's assistant, Benjamin Banneker, a black astronomer. *The Alexandria Gazette* noted that during the Masonic ceremony, corn, wine, and oil were laid upon the stone to symbolize nourishment, refreshment, and peace. The Rev. James Muir, pastor of the

Alexandria, Virginia

Location of the forty boundary markers is superimposed over Ellicott's map of the district.
Library of Congress, Geography and Maps Division

Aquia Stone Boundary Markers for the New Capital

This mural by William A. Smith is located at the Maryland House in Aberdeen, Maryland. It depicts Benjamin Banneker and Major Andrew Ellicott surveying Washington, D.C. for the placement of boundary markers.
State of Maryland, Department of Highways

Presbyterian Church of Alexandria, presented a prophetic oration about the infant nation. "From this stone," he said, "may a superstructure arise, whose glory, whose magnificence, whose stability, unequaled hitherto, shall astonish the world...."[3]

Andrew Ellicott, to sight his lines, had his men clean forty-foot swaths through forested land. Unfortunately, a number of his crew were killed by falling trees. This monumental undertaking of laying the Aquia stone markers required Ellicott's crew to trudge through forest wilderness, to plod through marshland, and to ford streams.[4]

Birthstone of the White House and Capitol

This photo shows one of the original forty boundary markers.
District of Columbia Public Library, Washingtonian Division

Each marker was one foot square and four feet tall. Two feet of the stone were embedded in the ground, leaving two feet exposed. (The four cornerstone markers were larger.) The top was beveled for four inches, creating a four-sided pyramid. The markers were apparently sawed and not chiseled, since some of the stones, as far back as 1905, still revealed saw marks.

On the side of each marker, facing the District of Columbia, was inscribed "Jurisdiction of the United States." On the opposite side was the name of the state it faced, either "Virginia" or "Maryland."

On one side was carved the year, and on the fourth side was, as Ellicott wrote, "the present position of the magnetic needle at the place."[5]

Major Ellicott wrote the Commissioners on January 1, 1793, "It is with singular satisfaction that I announce the completion of the four lines comprehending the Territory of Columbia." The boundary markers were placed approximately one mile apart, starting at Jones Point, "except as to a few places where the miles terminated on a declivity or in the water, in such cases the stones are placed on the nearest firm ground and the true distance in miles and poles is marked on them."[6] This monumental endeavor took two years for Ellicott and his men to accomplish.

Postscript:

In 1846, the district ceded back to Virginia the southern one-third. Thus today, on maps, the district does not look diamond-shaped. However, the boundary markers still lie in silent witness as to the district being "ten miles square."

Aquia Stone Boundary Markers for the New Capital

Prior to installing a protective fence, members of the D.C. Daughters of the American Revolution met at boundary marker NW4 in July 1915.
District of Columbia Society, Daughters of the American Revolution

In 1915, Members of the Daughters of the American Revolution wished to preserve the district's boundary markers. They raised funds and encased each boundary marker in a wrought-iron fence.[7] Today most markers are taken care of by DAR chapters in the District, Virginia, and Maryland.

In 2005, Gayle Harris, chair of the DC DAR Boundary Stone Committee, said only two of the original forty markers are gone. NE 1 was missing in the 1950s after a bulldozer cleared off land to build a strip mall. SE 2, near the present-day Alexandria train station, is a replacement stone.

3

Who Chose the Stone?

Most scholars write that George Washington, who made most decisions concerning the capital city, decided that the walls of the White House and Capitol be made of freestone. But nowhere is it documented. The writings of Washington, L'Enfant, and the Commissioners do not clearly state whose plan it was to use the Aquia sandstone. One must speculate as to how the stone was selected.

WAS IT GEORGE WASHINGTON?

George Washington, who was born at Wakefield in Westmoreland County, Virginia, moved to present-day Stafford County when he was six years old. His father wished to be closer to Accokeek Furnace, a large blast furnace that he owned in the county. George lived with his parents and siblings on a farm close by the Rappahannock River overlooking the town of Fredericksburg. When George was only eleven years old his father passed away. (At that time the farm was simply called "the Washington farm." Today it is known as "Ferry Farm," for there were several ferry landings on that site for ferry boats to cross the Rappahannock River.) After his father's death, George also spent time with his half-brothers at Wakefield and Mount Vernon.[1] But for the majority of his youth, he stayed in Stafford County until he was nineteen years old. Living in Stafford for thirteen years of his life, he was no doubt familiar with Aquia stone and Brent's Island.

Birthstone of the White House and Capitol

Portrait by Gilbert Stuart.
White House Historical Association/White House Collection

When George moved to Mount Vernon he did not forget the quarry near his boyhood home. In his ledger he wrote on April 28, 1774, that he paid Caleb Stone for his trip to "Acquia after stone for my steps." On May 3, 1774, he paid "Clemet Trig for 4 days hire of Mr. Clagget's flat…to bring stone from Acquia for steps."[2]

Even twelve years later, Washington's personal diary reflects the use of the stone:

> Sent my Stone Mason - Cornelius McDermott Roe, to the Proprietors of the Quarries of free Stone along down the River, to see if I could be supplied with enough of the proper kind to repair my Stone Steps & for other purposes. (Jan. 16, 1786)
>
> Cornelius McDermott Roe returned, having had the offer of Stone [from] Mr. Brent. (Jan. 17, 1786)[3]

Washington liked the look of stone. As a matter of fact, the wood on his own Mount Vernon was cut to look like stone. The process of painting the wood to resemble freestone was recorded by Washington himself:

> Sanding is designed to answer two purposes - durability, & representation of Stone…by dashing, as long as any will stick the Sand upon a coat of thick paint. This is the mode I pursued with the painting at this place…all my Houses have been sanded with the softest free stone pounded and sifted…. (He goes on to describe the sand particles that are to be thrown upon the wet paint.)…the fine dust must be seperated from the Sand by a gentle breeze, & the sifter must be of the fineness the sand is required….[4]

Although he does not talk about using freestone in the Capitol and President's House, he becomes very specific when writing to the Commissioners about walls with iron palisading in the new city. Evidently, the Commissioners were not specific about which stone they planned to use. Washington wrote, "I wish however you had declared that so much of the stone walls, on which the railing in the streets is to be placed, as shall appear above the pavement...should be of free stone hewed. The presumption I grant is, that no person who would go to the expense of an iron railing on a wall, would fix it on rough stone."[5]

Washington also presented the Commissioners with eight building regulations for the new federal city. The first was "...that the outer and party walls of all houses in the said City shall be built of brick or Stone."[6]

WAS IT L'ENFANT?

Evidently Washington expected the buildings to be grand, for he chose Major Pierre Charles L'Enfant to design the Federal City. (Inspired by the American Revolution, L'Enfant traveled to America in 1777 to become an officer in the Continental Army. Once here he Americanized his name to "Peter," which he used throughout his life.)[7] Actually Peter L'Enfant asked to be chosen for the task. He wanted to be employed to design the capital of "this vast empire."[8] Washington was familiar with his work, for the major had successfully remodeled the New York City Hall, where Washington took his presidential oath of office. At that time, the building was considered the most beautiful in America.

L'Enfant was eager to start planning the new city. He wrote, "No nation perhaps had ever before the opportunity offerd them to deliberately deciding on the spot where their Capital city should be fixed...."[9] It appears as though he wished to produce an original plan for the Federal City yet wanted to adapt it to the topography. He wrote Thomas Jefferson, the Secretary of State, and asked if he could send plans of Europe's great cities. He had no "Idea of Imitating"[10] the plans, but just wished to examine them for possible adaptation. The "grand cities" he asked for were London, Madrid, Paris, Amsterdam, Naples, Venice, Genoa, and Florence.

Jefferson searched for plans for twelve cities but found only two

The wood on Washington's home was cut to look like stone, then painted. Later, fine sand was cast on the wet paint.
Jane Conner (author)

of the ones L'Enfant requested—Paris, and Amsterdam. Enclosed with the plans he wrote, "I am happy that the President has left the planning of the Town in such good hands and have no doubt it will be done to general satisfaction."[11]

DID THE COMMISSIONERS CHOOSE THE STONE?

One can only speculate that Washington and L'Enfant together decided that the two most important Federal buildings be constructed out of stone. The Commissioners may have also had some influence. In October of 1791, even before the plans were drawn for the city and even before the competition for the design of the two buildings was held, the Commissioners requested that L'Enfant search along the lands near the Potomac for land well-stored with freestone. He returned from his journey after visiting various quarries and decided upon purchasing Brent's Island, which, after almost a century, was still owned and operated by Brents. George Brent owned the island, and Daniel Carroll Brent operated the island.[12] Most historians state that L'Enfant made his own decision regarding the purchase of the island. However, some say that he was ordered to purchase the Aquia Quarry. This might well have been the case, for one of the three Commissioners was Daniel Carroll, uncle of Daniel Carroll Brent. Besides that, the first mayor of Washington was another relative, Robert Brent.

On December 2, 1791, a deed was issued from George Brent (the fourth George Brent since the original George who had purchased it in 1694) to Peter Charles L'Enfant. It stated that L'Enfant and his successors were taking title of the island, except one square acre, for the benefit of the public. The island was sold for the sum of 1.800 pounds Virginia currency or $5,400. Two months later George Brent granted and confirmed title for the island to Trustees for the Commissioners, "establishing the temporary and permanent seat of the United States."[13]

The one-acre parcel, mentioned in the deed, belonged to a stone mason and cutter from Baltimore by the name of Robert Steuart (variously written as Stewart or Stuart). Steuart had purchased the acre of land five years before from George Brent for 50 pounds in Virginia currency. The acre was defined by four stones marked with the owner's initials, "R.S." In this deed the island was "called and known by the name of Brent's Island."[14] (Today two of the stones are still visible upon the island.) Documents do not reveal why Robert Steuart did not sell his acre to the Federal Government. Perhaps he thought he could make more money by selling his stone in Baltimore.

L'Enfant also arranged for the rental of another quarry on Aquia Creek from John Gibson, a businessman from Dumfries, Virginia. John Gibson bought the adjoining tract from the Brent family shortly before the government transaction. The exact location of his quarry was not mentioned in the Commissioners' records, but was thought to

Who Chose the Stone?

be close to the island, for a canal was to be dug connecting the two quarrying facilities. The rental was to be for ten years at sixteen pounds Virginia currency per year with an original twenty pounds rent to cover "all the land which you have adjoining to the quarry."[15] The rental agreement was terminated after six years. The Commissioners wrote Gibson, "We are nearly done with [your] free stone" and wanted him to "take the quarries off our hands."[16]

After the purchase of the island by L'Enfant, rumors were circulating around Washington, D.C., that George Brent had found out from his relative, Commissioner Daniel Carroll, exactly how much the Commissioners were willing to pay for the island quarries. Thus, when L'Enfant traveled down to Stafford County to negotiate the deal, he was unable to obtain the island at a lower price because Brent had been forewarned. The Commissioners believed that George Walker, a Scottish Georgetown merchant, had spread these rumors, so they wrote a letter to him asking that he reveal where he obtained such information. Walker wrote back that it was "Major L'Enfant or one of the company" at Suter's Tavern or Davidson's Counting Room. Walker said if they did not believe him, he would be ready for a duel.

Commissioner Daniel Carroll refuted Walker's claim and said that he never informed his Aquia relatives. He even sent an affidavit to James Madison in which George Brent was quoted as saying, "I never received from Mr. Carroll either directly or indirectly or from any other person the smallest intimation of the price the Commissioners were disposed to give."[17]

It remains unclear as to who selected the Aquia freestone—Washington, L'Enfant, or the Commissioners. The decision, however, had practical significance, as freestone could be acquired nearby and transported down Aquia Creek and up the Potomac, thus avoiding delays of waiting for costly imported stone. Also, it was appropriate that the new nation contribute its native resources for the construction of the Federal City's lasting structures.

L'Enfant purchased all but one acre of the island. It belonged to Robert Steuart who marked it with four boundary stones. The largest is visible on the island today.
Lou Cordero
Fredericksburg Free Lance-Star

4

L'Enfant and the City's Genesis

The year 1791 was a very hectic one for the new capital city. Andrew Ellicott and his men were busy clearing land and placing boundary markers. L'Enfant was drawing the plans for the city that he thought should be "on a grand scale." He selected an area known as Jenkins Hill for the Capitol or "Federal House." He described the wooded hill "as a pedestal waiting for a monument." The "Palace for the President" would be at a site already suggested by Washington. L'Enfant wanted the building to have "the sumptuousness of a palace, the convenience of a house and the agreeableness of a country seat."[1]

Thomas Jefferson suggested laying out the city using a plain rectangular plan such as in Philadelphia, but L'Enfant wanted to take advantage of the area's topography and river system. He wished to create a city that was "grand and truly beautiful." His vision included streets emanating from radiating axes, ever-flowing fountains, lavish gardens, monuments, and squares with statues, columns, and obelisks dedicated to each founding state of the union. He even wanted to have a cascading waterfall flowing from under the base of the Capitol. Avenues would be wide, but the grand avenue, he wrote, "connecting both the Palace and the Federal House will be most magnificent and most convenient." L'Enfant was insistent that the grandeur of the city should be equal with the greatness of the United States in future times.[2]

Birthstone of the White House and Capitol

L'Enfant by Bryan Leister.
Historical Society of Washington, D.C.

George Washington was eager to inspect every detail. Even though he was knowledgeable in surveying and had given instructions for construction at Mount Vernon, he evidently did not feel qualified in the field of architecture. Once he wrote, "…I profess to have no knowledge in Architecture, and think we should (to avoid criticisms) be governed by the established rules which are laid down by the professors of this Art."[3] Regardless of how he felt, Washington's final approval was given to every aspect of the construction of the new Federal City.

In July, the clearing of the forests began, brick kilns were erected, and foundations were dug for the President's Mansion and Capitol.[4] In September the Commissioners selected a name for the district and the city. They wrote to L'Enfant, "We have agreed that the federal district shall be called 'the Territory of Columbia,' and the federal city 'the City of Washington.'" It is interesting to note that in practically all correspondence and conversations, Washington was too modest to call the city by his name. He always wrote such things as "the Federal City." In his will, however, he designated the city with its proper name for legal purposes.[5]

By the end of 1791, L'Enfant wanted to push forward work during the winter, so he wrote to his assistant, Issac Roberdeau:

> …I have to recommend to your particular care the following…. To repair immediately to Acquia Creek to see the qurries [*sic*] there belonging to the public - to have barracks erected theron for twenty men on each of these quarries, on the island purchased from Mr. George Brent and on that rented from Mr. John Gibson of Dumphries…. The exporting of stone must be begun at once on both quarries; they must be opened at once all around the island and on the main [land], on the whole front adjoining the creek. The stone must be taken down as it comes and any size and in as great quantity as the time will admit, recommending only that when the rock will be pound sound and free from staion that blocks of stone be extracted therefrom the largest size every way as is possible. When arrived at Philadelphia I shall send you the particular dimensions of some stone-but without waiting let the hands do the most they can-when the weather shall prove too severe let them busy themselves in clearing away the rubble and as soon as often as it moderates let them set about extracting the stone.[6]

On Christmas Day, L'Enfant prepared another trip to Philadelphia. This time he traveled there to make ready the plan of the Federal City for engraving. He left Roberdeau in charge of operations and instructed him to take twenty-five of the men engaged in the city work and go to the stone quarries at Aquia Creek. The Commissioners thought it was more important that the work of digging up clay for bricks be pushed ahead and requested him not to travel to the quarries. Instead, the faithful Roberdeau proceeded to Aquia Creek and was arrested by the Commissioners. L'Enfant was furious. This incident was just one of the many rifts that occurred between the planner and the Commissioners during this planning period.[7]

By January of 1792, L'Enfant prepared a list of the things that had to be done to commence and facilitate construction of the two principal buildings. Some items on the list pertained to the preparation for the freestone's arrival. For example, he estimated that for four months, 300 men would be required to build a canal to transport stone from the bank of the Potomac to the Federal Square. Afterwards, the men would be hired to work at either end of the canal.

Also, two large scows "of particular construction" would be needed to transport large stone and two smaller scows to transport "smaller stones." They "must be constantly employed and will require twenty boatmen."

Twenty stone cutters with ten assistant laborers would be required initially. However, around July 4th, when the "materials are collected in sufficent quantity round the buildings…the twenty masons must be increased to 40 with the addition of 60 labourers."[8]

Unfortunately, L'Enfant himself was never able to put all his plans into action. Just two months after his plans were presented, the Commissioners informed him "We have been notified that we are no longer to consider you as engaged in the business of the federal city."[9] L'Enfant had too many arguments with those around him—especially the Commissioners. He became a lonely, poor man. Benjamin Henry Latrobe, who became Architect of the Capitol, described the later years of L'Enfant's life. "Daily through the city stalks the picture of famine, L'Enfant and his dog…. He is too proud to receive any assistance and is very doubtful in what manner he subsists."[10] This is in striking contrast to a description of L'Enfant before. L'Enfant "was a tall, erect man, fully six feet in height, finely proportioned, nose prominent, of military bearing, courtly air, and polite manners, his figure usually enveloped in a long overcoat and surmounted by a bell-crowned hat - a man who would attract attention in any assembly."[11] Today, 200 years later, L'Enfant's name graces streets, squares, and buildings paying homage to this creative city planner. However, L'Enfant went to his grave never receiving recognition or adequate pay for his accomplishments.

5

Jefferson, Brick vs. Stone

Thomas Jefferson, the Secretary of State during George Washington's first administration, had a close working relationship with the President and the Commissioners. He recognized that Washington held the authority, yet seemed secure in offering suggestions. If Jefferson had his way, the "Federal City" would look nothing as it does today, for he preferred buildings to be made of red brick with stone trim, much like his own Monticello. Once he wrote the Commissioners, "The remains of antiquity in Europe prove brick more durable than stone. The Roman brick appears in these remains to have been 22 inches long, 11 inches wide and 2 inches or 2 ½ inches thick." He even wrote that Washington preferred brick with stone trim rather than stone. "On speaking with the President on Mr. Stewart of Baltimore's idea of facing the buildings with stone of different colors, he seemed rather to question whether from the watertable, perhaps from the ground upwards, brick facings with stone ornaments would not have a better effect, but he does not decide this."[1]

In March of 1792, Jefferson prepared two advertisements for a contest to design the Capitol and President's House. After the approval of Washington and the Commissioners, the advertisements went out to major newspapers around the country. The advertisement for the Capitol said that whoever produced the most approved plan would receive $500 or a medal.

Birthstone of the White House and Capitol

Thomas Jefferson by Rembrandt Peale.
White House Historical Association/White House Collection

The ad clearly stated, "The building to be of brick...." The advertisement for the President's House also gave the same prize, but required the contestant to give "an Estimate of the Cubic feet of brickwork composing the whole mass of the walls."[2] (This raises a curious image of a Red House rather than the now familiar White House.) Washington was given the opportunity to change the advertisements. As a matter of fact, he did make comments with pencil in the margins. One would have thought he would have scratched out "brick," considering that L'Enfant had already purchased the Aquia quarries, but he did not.

Even though the advertisements were printed, Washington started his own search for an architect. The year before, while visiting Charleston, South Carolina, he heard about the work of a Dublin-trained architect, James Hoban. Once the contest began he gave Hoban a letter of introduction for the Commissioners. The letter stated, "…I have no knowledge of the Man [Hoban] or his talents, further than the information which I received from the Gentlemen of Carolina...."[3] In the meantime, entries were coming into Philadelphia, the then Federal City, for the inspection and approval of the President. A disappointed Washington wrote, "If none more elegant than these should appear..., the exhibition will be a dull one indeed."[4]

The day after the competition closed, on July 16, 1792, Washington arrived in Georgetown for the final judging. Hoban's entry was chosen. Ignoring Jefferson's advertisement and the request for a brick building, Hoban presented a plan for a stone house. (Jefferson, himself, even submitted an entry. He signed the work with the cryptic pseudonym "A. Z.") Hoban's entry was reminiscent of the Leinster House in

John C. Rauschner created this wax portrait of James Hoban.
White House Historical Association/ White House Collection

Hoban, an Irish architect, won the design competition for the President's House and selected a gold medal and the remainder of a $500 prize in cash.
Maryland Historical Society

William Thornton.
Architect of the Capitol

Dublin, Ireland. The Leinster House was a ten-year-old stone building that was the residence of the Duke of Leinster.[5]

Unfortunately, there was no winning entry for the Capitol. There were few professional architects at the time, so most entries were not satisfactory. Fortunately, a physician by the name of William Thornton received permission to submit a belated entry. Thomas Jefferson was delighted by Thornton's design and said it "captivated the eyes and judgment of all." George Washington was so pleased that he wrote the Commissioners he had no doubt that they would approve Dr. Thornton's design, for it combined "grandeur, simplicity and beauty."[6]

Aquia stone was also used for something other than the construction of the district's two most important structures. Jefferson, and a gentleman by the name of William Lambert, were convinced that this new independent nation should not rely upon the mother country's prime meridian running through Greenwich, England. Instead, they felt that American cartographers and navigators should have the prime meridian running through Washington, D.C.[7] Such a survey did take place, and a meridian was established on September 20, 1793. Eleven years later, a report was given. It stated that "the first Meridian of the United States" intersects the center of the north and south basement doors of the President's House. Several spots were marked with temporary posts. Later,

William Thornton won the competition for designing the United States Capitol. Trained as a physician, Dr. Thornton entered this contest as an amateur architect. He won $500 and a city lot.
Library of Congress

Aquia freestone was used to delineate these specific locations. One marker was located by the present-day Washington Monument in direct north-south line with the White House. The other was an obelisk that was called either Jefferson Obelisk or Meridian Pier. According to a Commissioner of Public Buildings who in 1889 dug up the foundation of the old pier, "The pier was frequently used by surveyors as a bench-mark, and as a guy-post for barges and other boats."[8] (By 1889 the freestone markers had disintegrated, as no one had kept them in repair. Today other stones mark the location of America's attempt to establish prime meridian.)

6

Cornerstone Ceremonies

WHITE HOUSE CORNERSTONE

Congress passed the Residence Act in 1790, which stated that the American Government would move into the new Federal buildings within ten years. By 1792, the planners had eight years left. They had to work rapidly now that the designs for the buildings were selected. The Commissioners had tried to raise money by selling lots in the district. The first effort was not too successful, so it was thought that if they held the cornerstone ceremony for the President's House around the same time as the second sale of lots, they would show the nation that it was indeed a wise investment to purchase an interest in the new Federal City.

The cornerstone celebration was held on October 13, 1792. A group of dignitaries first assembled at the Fountain Inn in Georgetown. President Washington could not attend, as he was in Philadelphia. So the parade consisted of the three Commissioners; James Hoban, the contest winner; Collen Williamson, the master stone mason; and Free Masons. They marched from the inn to the site.

The parade participants were probably greeted by a scene of workers' huts, pits for slacking lime, and trenches filled with foundation stone. Previously, L'Enfant had Benjamin Banneker mark off the location for his palace-sized structure with wooden stakes. Now the new house, designed by Hoban, would be much smaller. Realizing the new design dwarfed L'Enfant's, the Commissioners asked Washington to determine where the house should be located. After much calculation, he determined the house would face north just as L'Enfant had planned. After direction from Washington, Hoban supervised the laying of the foundation.[1]

Foundation stone was in abundance. It could be purchased and transported easily from quarries close to, or in, the Federal City. Foundation stone was not Aquia stone, but rather any common stone such as pieces of schist, gneiss, or granite. A name used for such stone was "Potomac Bluestone" since there were outcroppings of such rocks along the Potomac River. One outcropping was just north of the present Lincoln Memorial on a promontory that served as a surveyor's bench mark. Old maps referred to it as Key of Keys. Later, after British General Braddock and his soldiers landed there in 1755, it was called Braddock's Rock. Records show that a quarry was established there for the express purpose of obtaining bluestone.[2] Other foundation stone was obtained from several quarries near the construction sites, like Rock Creek.

A thick layer of foundation rubble stone was placed on a bed of clay. Next, rough-cut stones from Aquia quarries were laid. Dressed, rectangular-cut Aquia freestone blocks would later be placed, starting at ground level. The area was ready for the cornerstone ceremonies. The cornerstone was probably Aquia stone, but historic documents do not disclose the type of stone. The use of freestone is assumed since foundation stone was usually quarried in smaller blocks and could not be dressed appropriately for ceremonial use.

A Masonic ceremony took place with an oration delivered by the master of the Georgetown lodge. On the southwest corner the cornerstone was laid. A Charleston, South Carolina, newspaper was the only one to print a detailed report. It read:

> Under the stone was laid a plate of polished brass, with the following inscription:
>
> "This first Stone of the President's House was laid the 13th Day of October, 1792, and in the 17th Year of Independence of the United States of America.
>
> George Washington, *President.*
> Thomas Johnson,
> Doctor Stewart,
> Daniel Carroll,
>   *Commissioners*
> James Hoban, *Architect.*
> Collen Williamson, *Master-Mason.*
>   Vivat Republica."
>
> After the ceremony was performed they returned, in regular order, to Mr. Suter's Fountain Inn, where an elegant dinner was provided [and enumerated toasts were given].[3]

CAPITOL CORNERSTONE

Eight months after the cornerstone laying at the President's House, the Commissioners received information about the Aquia quarries from James Hoban and Collen Williamson. Since a large quantity of stone would be needed for the winter months at both the President's House and Capitol sites, they expressed a desire that "…a canal may be cut about 18 feet wide to let scows in to the quarry on the Island - Patrick Whelan has been talked with and probably may execute it."

At the same meeting, in June of 1793, the Commissioners discussed the proposed Capitol. "As soon as the president's sentiments are known…a corresponding foundation to be dugg and preparation made for laying that Foundation." A Mr. Williams, who provided foundation stone, was to be contracted. Stone cutters were to be retained. "They will for that purpose be employed on the Foundation of the Capitol and on the Hotel while the stone is accumulating from the quarries." The work force at the quarries was also to be increased.[4]

According to Bill Allen, Architectural Historian of the Capitol, a process called "continental trench" was used to create the foundations. A ditch was first dug. Then a wheelbarrow of various rubble rocks was dumped into the trench. Next, a wheelbarrow of mortar was poured. This process was repeated until the foundation reached the proper height. Today if one examines the Capitol's original foundations it is evident that this haphazard method was used.[5]

Benjamin Henry Latrobe depicted President Washington and fellow Masons marching to the Capitol site on September 18, 1793, to lay the cornerstone. *Alexandria-Washington Lodge, No. 22 AF&AM, Alexandria, Virginia*

Birthstone of the White House and Capitol

At a September 2, 1793, meeting of the Commissioners, it was stated that the work on the foundation of the Capitol was in progress, but a southeast corner was kept vacant for the cornerstone. A ceremony would be held on September 18 in which masons from "both sides of Potomack" would attend. A request was published that, "Those of the Craft, however dispersed, are requested to join the work. The Solemnity is expected to equal the occasion."[6]

Unlike the Presidential Mansion cornerstone festivities, President Washington was able to participate in the activities for the new Capitol building. On September 18, 1793, approximately two weeks after the advertisement appeared in various newspapers, Washington left his Mount Vernon home and traveled to the southern shores of the Potomac River, where he was greeted at 10 o'clock by a procession of Freemasons and a company of Volunteer Artillery. The Artillery, according to the *Columbian Centinel*, "…paid their military honors, and his Excellency and suite crossed the Patowmack…." Once on the Maryland shore they were greeted by more Masonic lodges, stone cutters, mechanics, and a band. A parade was formed with the President in the rear. "The procession marched two abreast, in the greatest solemn dignity, with music playing, drums beating, colours flying and spectators [of both sexes] rejoicing."[7] At the President's Square they reversed the order of their march. The procession moved in single file across "the rude bridge formed of a single log"[8] and went to the southeast corner of the Capitol site.

On October 23, 1993, there was another parade and cornerstone reenactment for the 200th anniversary of the "Laying of the Cornerstone."
Author

A flowery oration was delivered by the Masonic grand master. He marveled how much masons from two states had accomplished thus far and talked about how much could be achieved when all fifteen states sent representatives of the craft. They would be contributing "an universality of individuals, like innumerable hives of bees bestowing their industrious labor on this second paradise."[9] Then Washington and ranking members of the Free Masons took "their stand to the east of an huge stone; and all the craft forming a circle, westward, stood a short time in silent awful order. The Artillery discharged a volley."[10]

The Masonic cornerstone ceremonies for the Capitol were just like those for the district's boundary stones and President's Mansion. However, an engraved silver plate was used rather than one of polished brass. President Washington "desposed the plate, and laid on it the Corner Stone...." Corn, wine, and oil were poured on symbolizing nourishment, refreshment, and peace. (The silver trowel used by Washington during the cornerstone ceremony is kept at a museum in the George Washington Masonic National Memorial in Alexandria, while his marble-headed gavel is in possession of Potomac Lodge #5 of Georgetown. They are taken out of storage and used only for special occasions. For example, U.S. presidents have used them in the dedication of public buildings. They were used in the reenactment celebrations for the 200th anniversary of both the White House and Capitol. For the Capitol's original ceremony in 1793, Washington may have worn the ceremonial Masonic apron embroidered for him by the Marquis' wife, Madame LaFayette, or one given to him by a Masonic brother. Both aprons are the property of Masonic lodges.)[11]

The ceremonies ended with prayer and a fifteen-volley salute from the artillery, one for each state in the Union. *The Columbian Centinel* article closed by saying, "The whole company retired to an extensive booth, where an ox of 500 lbs. weight was barbecued.... Before dark the whole company departed, with joyful hope of the production of their labour."[12]

Work continued on both structures after the celebrations. George Washington's private secretary wrote about the progress of the structures and referred to the Aquia freestone in a book published in 1793. "The public buildings...are on a scale equal to the magnitude of the objects for which they are preparing; and will...be executed in a stile [sic] of architecture, chase, magnificent and beautiful. They will be built with beautiful white stone...."[13]

7

Many Workers Are Needed: Slave, Free, and Skilled

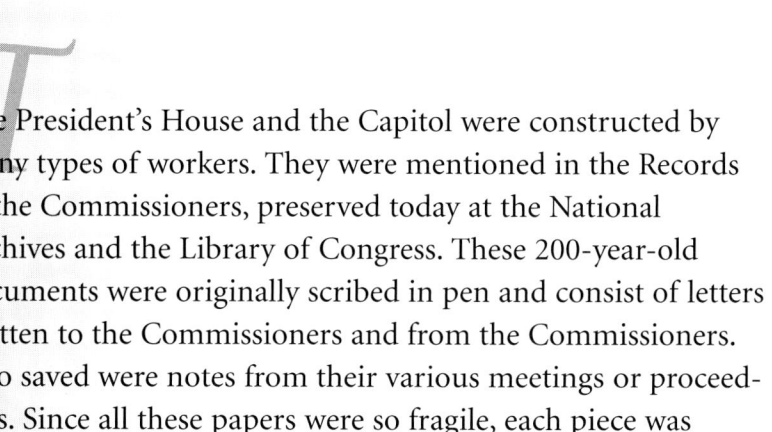

The President's House and the Capitol were constructed by many types of workers. They were mentioned in the Records of the Commissioners, preserved today at the National Archives and the Library of Congress. These 200-year-old documents were originally scribed in pen and consist of letters written to the Commissioners and from the Commissioners. Also saved were notes from their various meetings or proceedings. Since all these papers were so fragile, each piece was photographed, placed on microfilm, and catalogued according to date. Also at the National Archives are treasury records that include such things as pay vouchers and bills of lading. The study of all of these documents, as well as presidential correspondence and newspaper advertisements, both inform and enlighten as to the classes and types of workers.

The Commissioners were gentlemen, men of breeding, landowners with formal education. Their records were detailed. Few workers, however, wrote letters. Most signed only their names when requesting such things as increases in wages or when stating their grievances. For the most part, they did not indulge in corresponding or in the writing of diaries. Most common laborers and slaves were illiterate and kept no records whatsoever. Therefore, the records of the

Commissioners are an invaluable tool to identify the type of workers who built these two impressive structures.

CLASSIFICATION OF WORKERS

Each worker had a unique responsibility and was needed to complete the task of building the nation's most significant structures.

> QUARRIERS removed the blocks from the quarry. The blocks they quarried were roughly shaped and still contained pick marks.
> STONE CUTTERS cut the stone into desired sizes.
> STONE SETTERS placed the stone in position at the work site.
> ROUGH-MASONS finished the blocks or "DRESSED" the stone. They sharpened the edges and smoothed the sides.
> STONE CARVERS cut moldings and carved intricate details.
> MASTER-MASONS were usually overseers over the entire stonework operation.[1]

Master-Masons were considered close to the Commissioners in class. From their letters to the Commissioners, it appears that they considered themselves on approximately the same footing. However, when they wrote to the Presidents, they did show proper respect.

MASONS

Today when a person states that his occupation is that of being a "mason," we think that he is one who works with brick, concrete, or stone. But during the colonial period, a mason was one who worked *only* with stone. A freemason was one who worked primarily with freestone.[2]

There are no known original drawings of workers at the quarries. This drawing by William Strickland shows a carver chiseling furrows into stone. Around him are tools commonly used: workbench, square, level, hammer, pick, straight edge, dividers, trowel, and frame saw. Column capitals, bases, moldings, and mortuary work surround him.
Ewell Sale Stewart Library
The Academy of Natural Sciences of Philadelphia

Just as in today's society, there were also Masons or Freemasons who belonged to a fraternal order. Originally, Freemasons formed guilds that were restricted to stone cutters. However, in England, after the Reformation, they admitted men that had great social status or wealth. In 1717 a Grand Lodge was formed in London from which all other lodges around the world were derived. In 1733 two lodges were founded in America, one in Boston and one in Philadelphia.[3] At the time of the American Revolution, there were 150 lodges in the colonies. George Washington was a Mason as were a large percentage of other signers of the Declaration of Independence. Virtually all building construction in the new Federal City proceeded only after Masonic cornerstone ceremonies were held.

Freemasonry held ideals of religious toleration and basic equality of all people. They traced their beginnings to the building of King Solomon's temple in 1030 B.C.

The new Federal City was going to have its own temple. The proposed Capitol was to be the largest and most important symbol of freedom for the nation. Jefferson wrote that it was "the first temple dedicated to the sovereignty of the people."[4]

SKILLED WORKERS

The skilled workers included artisans, stone cutters, stone carvers, and blacksmiths. These skilled workers owned their own tools and had already gone through their apprenticeships. A young man who wished to be an apprentice usually started his training anywhere from fourteen to seventeen years of age. His family would pay a certain amount to a skilled mason in order for him to become proficient in the trade. By the time the son was twenty-one, he would be considered a journeyman and would work for a master craftsman. The total number of years of apprenticeship was from three to seven.[5]

Obtaining skilled mechanics or workmen was of great concern to the Commissioners. At first they looked to the eastern states for stone cutters, brick layers, carpenters and masons. When it was found that these positions were going to be hard to fill, the Commissioners tried to obtain foreign labor. They advertised in both Scotland and Holland and stated that they would pay travel expenses. Even this failed to bring adequate numbers, so the Commissioners decided to make their offer more attractive. They agreed to advance passage money, to provide transportation for the men's wives and give assurances of their social standing upon arrival. The skilled workers were eventually obtained from the eastern United States and from abroad—mostly Scotland, Holland, Germany, and France.[6]

SLAVE and UNSKILLED LABORERS

Common labor was not hard to obtain since many residents of both Virginia and Maryland were eager to have the opportunity of hiring out their slaves.[7] In the spring of 1792, the Commissioners, who were slave owners themselves, wrote that they would "hire good labouring negroes by the year, the masters cloathing them well and finding each a blanket, the commissioners finding them provisions and paying twenty one pounds a year," or $55 to their masters.[8] The government provided food and shelter. In Washington, D.C., a hospital was erected for the common laborers.[9]

In the late 1790s a Polish literary figure and patriot, Julian Niemcewicz, visited Washington and was shocked to see the conditions of the slaves at the Capitol. In his diary he wrote:

> It was eleven o'clock. No one was at work; they had gone to drink grog. This is what they do twice a day, as well as dinner and breakfast. All that makes four or five hours of relaxation…. The negroes alone work. I have seen them in large numbers and I was very glad that these poor unfortunates earned eight to ten dollars per week. My joy was not long lived: I am told that they were not working for themselves; their masters hire them out and retain all the money for themselves. What humanity! What a country of liberty. If at least they shared the earnings![10]

The largest foreign-born population in the district in the early years was that of the Irish. In 1796, it was recorded that almost every year a vessel filled with Irish laborers arrived at Georgetown. They competed with slave labor and obtained jobs as ditch diggers, brick makers and stone cutters. They also dug canals and laid out streets.[11]

White laboring classes could be found living in shacks and shanties scattered in practically every part of the city. However, there were two main settlements. One was close to the Navy Yard, and the other was located in Georgetown. In 1800 Secretary Walcott spoke of "small, miserable huts, which present an awful contrast to the public buildings."[12]

HOW MANY WORKERS?

It is difficult to say exactly how many workers were at the quarries and at the building sites, as the quantity changed according to assignment or weather. Pay records, reports to the Commissioners, and diaries give some indication. For example, in the winter of 1794, Hoban gave a report to the Commissioners regarding the state of the buildings and an estimation of how many workers would be needed the following year. In it he stated that twenty stone cutters were working on the President's House and twenty would be needed to work in 1795 along with thirteen laborers. He also stated that at the Capitol twenty stone cutters were employed.[13]

Another example of determining the work force is from firsthand accounts. Anne Newport Royall, upon viewing the Capitol in 1825, wrote:

> The capital lacks a great deal of being finished; although a number of hands (200) are constantly at work upon it, it is thought it will take twenty years to complete it. It would astonish any one to see the immensity of stone lying about it, (one would think enough to build another capitol.) which remains to be put somewhere, but it would puzzle Apollo to tell where.[14]

In the year 2000, Ed Hoatling, an NBC news producer, was doing research for a piece about the 200th anniversary of the U.S. Capitol. While poring over Treasury Department pay slips, he discovered that out of 650 workers at the Capitol, from 1792–1800, 50 were free blacks, and up to 400 were slaves.

LODGING, WAGES, AND FOOD

Housing for workers consisted of wood or brick houses. Some were located near the President's House, by the present-day Lafayette Park, while other housing was located near the Capitol site. A description of the brick houses can be found in the Commissioner's Records of March 14, 1793:

> …it will be best to erect brick houses of two stories for the accommodation of the workmen divided into rooms of about 10 feet square on public lots near the President's house, two rooms to one Stack of chimneys and four rooms centered on the division line of two lots.[15]

This illustration by Charles Tomlinson depicts early-nineteenth-century masons cutting and dressing stone.
Library of Congress

Quarters for men were built on Government Island. The first mention of quarters were those that L'Enfant requested.[16] Records do not indicate if they were ever constructed under his guidelines. However, the next mention of lodging was from the Commissioners on March 28, 1792. They entered into an agreement with George Brent to erect four huts on Wiggington's Island to be completed by April at a cost of seven pounds ten shillings each.[17] Later records indicate that quarters on the island cost approximately $400.[18] Foundations of a building approximately fifteen by thirty feet can be found on the island today. One does not know whether skilled laborers and common workers all stayed on the island together. Perhaps slaves returned to their masters' homes after a day's work.

A full day's work was usually from sunrise to sunset, thus making a longer day in the summer months. The skilled mechanics worked from 6 a.m. to 6 p.m. until 1814, when they worked ten hours. Many times work ceased altogether if the weather was bad.

Niemcewicz, the Polish visitor, observed the skilled workers in May 1799. He stated that they commenced work at 6 o'clock but stopped between 8 and 9 for an hour breakfast. Dinner started at one and also lasted an hour. During his visit to Washington he also noted the following:

> I have seen them often quit their work, come into the little dramshop in order to talk while drinking a glass of *grog*. Once I went into the hut of one of these workers. I found his wife there dressed very neatly, good utensils for cooking, and all the service for tea in porcelain from China. Far from being scandalized by this small luxury, I rejoiced in it. Why should not a man who works by the sweat of his brow enjoy the comforts and the ease of life. Should it be only idlers who have these privileges?[19]

Pay records show that there were normally two holidays, those being Independence Day and Christmas. The Christmas vacation lasted three days.[20] Masons also celebrated the Festival of Saint John the Baptist. They invited the "community of the craft" and usually walked in procession from a lodge to the Capitol site or a church. Services were held, and then the procession went to a hotel or another location for refreshments. Money received from the sale of tickets went to "relieve distressed widows and orphans of deceased Brethren…."[21]

The laborers' wages were 75 cents a day, while skilled stone cutters received from $1.25 to $1.75 per day. The overseer of the laborers received $2.12 while the overseer of the stone cutters received the highest wage of $3.75. Immediately after the War of 1812, wages were almost doubled, but three years later they returned to the previous lower amounts.[22]

The food for the workmen was decided upon by the Commissioners in April of 1792. "The men to be found provisions by the Commrs., that is one pound good pork or one pound and a half of beef and one pound of flour per day all days included."[23] A quarry account book for a much later period, 1838, shows that the owners of an Aquia quarry bought the following supplies for their workers every month:

> 75 lbs Bacon
> ½ Bushel meal from Chelsea
> 125 Herring
> 1¼ gallon whiskey
> 3 gall 1 pt. molasses[24]

Other account records from the same quarry indicate that items such as "Irish" potatoes, beans, corn, and tobacco were brought to the quarry, too.[25] Undoubtedly, similar type goods were also delivered fifty years before to Aquia quarry sites to supplement the flour, pork, and beef.

Liquor was also included in their diet. After the cornerstone ceremony at the President's House, the Commissioners had difficulty getting an adequate supply of stone. In order to speed up production, the Commissioners passed an order that whiskey be provided at the quarries. A half-pint a day was to be delivered to each man. They later said that the supervisor of the quarry, at his discretion, should increase the daily ration in extraordinary cases.[26]

However, in 1793 the Commissioners tried to regulate the sale of liquor for workers in the city as well as those on the island. Robert Brent, manager of some Aquia quarries, made the following agreement:

> I further bind myself not to keep Liquor for sale, nor permit it to be sold by any person on the said tenement so as to interrupt the public work in the quarries.[27]

Evidently this did not last long, as liquor was considered an indispensable commodity. A few months later, upon Master Mason Collen Williamson's recommendation, "…the Commissioners order that a quantity of whiskey be provided at the Stone Quarries and half a pint a day delivered to each man employed there till the 15 of Septr next."[28]

A year later, in 1794, the Commissioners approved increasing the liquor since the sickly season was approaching. Apparently the commissary at the Aquia quarry took them up on their offer, for it purchased thirty-one gallons of whiskey for the thirty-five men employed on the island![29]

Later, when the walls of the President's House were quite tall and scaffolding was needed, the amount of liquor was increased. William Seale writes in his two-volume set, *The President's House*, "In the hot summer, men who worked on the high scaffolding at the house were given a pint of whiskey or rum each day, with an extra half-pint in the burning days of late August and September, also to build resistance and encourage strong bodies."[30]

Besides giving lodging, food, and whiskey to the workers, it appears as if the quarry workers also received "quarry cloathes" and shoes.[31] One Aquia quarry's account book indicates that a woman was paid for making quarry pants, shirts, and roundabouts.[32] (A roundabout is a short, closely fitting jacket or coat.) These items of clothing do not appear in the Commissioners' papers or account records, only in quarry records. Therefore, it appears as if clothing was supplied by individual quarry managers or obtained from the slaves' masters.

RELIGIOUS HERITAGE OF WORKERS

Workers from abroad brought with them their culture and religion. The first churches in the district were the result of workers worshiping on Sunday, their one day off. For example, the first Presbyterian church, within the city of Washington, consisted of stone masons, mostly Scottish, who met in the carpenters' shop near the grounds of the President's House in the fall of 1793. Later, in 1800, they worshiped in the lobby of the Treasury Building.[33] After the Presbyterians, the Methodists and Baptists also used the carpenters' shop. With Hoban's assistance, the St. Patrick's Catholic Church was founded in the 1790s. The parishioners met in a wooden house a few blocks from the site of the President's House.[34] A converted tobacco warehouse on New Jersey Avenue served as a place of worship for Episcopalians. The Society of Friends built a meetinghouse in 1808 on I Street. The chamber of the House of Representatives even served as a place of worship for various denominations. Visiting ministers would preach from the Speaker's desk, while a uniformed Marine band would play "pieces of psalmody."[35]

8

Quarrying Aquia Stone

Stone orders submitted to the Aquia Quarries specified specific types of stone to be used in the new federal structures.

Rubble was normally used for foundations and included small pieces of broken stone that had been scattered throughout the quarry, or rough fragments of debris chipped away to form trenches or work areas.

Scabble was roughly dressed stone in which pick marks could be seen upon its surface. Blocks were usually cut to a uniform size to reduce shipping weight. A pick was often called the "scabbling pick."[1]

Ashlar was a hewn or squared stone. Usually it was a thinner stone, approximately one foot thick, that would face a wall of brick.

Bill was stone cut to a specific dimension for a specific area. For example, if the master mason, working on the President's Mansion, desired a piece of stone for the southeast wall by a window, he would give the exact dimensions. At the quarry, the stone might even be coded SE or given a certain number. When it arrived at the site in the district, the masons would know exactly where it should be placed after final dressing.[2] "Bill" stone might also be specific stone cut for moldings, cornices, corners, and friezes. Today some stones still can be found on the island that appear to be cut to early specifications.

Quarrying the stone was a difficult task. First, all vegetation had to be removed. Next, layers that might contain roots of trees were cleared away to expose good stone. Once the top was cleared, the facing, or vertical stone, had to be picked away, creating a clear working area. After that, channels, or trenches, were dug into the wall, usually twenty feet apart. The remainder of the process is described by Lee Nelson in his book, *White House Stone Carving*.

> …trenches were only about twenty inches wide, providing barely enough room for a man to work with a pick and cut a relatively smooth surface on each side of the trench. Then a rear trench was cut behind and parallel to the initial stone face, and it connected the two side trenches. This last trench effectively created a very large rectangular mass of stone that could be split into manageable sizes.
>
> The quarriers then chiseled shallow horizontal and vertical grooves one to two inches wide between the trenches in the face of the stone. These grooves provided a plane from which stone blocks could be wedged away from the main mass of stone. The location of these grooves or cutting planes depended upon the presence of veins or other flaws within the stone itself, as well as the specific size of stone needed…. To split the stone away, a number of iron wedges were placed into the grooves about one foot apart and systematically and uniformly driven into the grooves, splitting the large block into the desired size.[3]

A piece of bill stone, thirty-six inches in diameter with a thickness of four inches, rests on Government Island. It appears to be awaiting shipment to D.C.
Author

A blacksmith was an indispensable worker at a quarry or building site. He had to sharpen or make tools constantly. For example, on March 30, 1792, the Commissioners ordered "twenty pecks and ten [1 peck = 8 quarts, dry measurement] trimming hammers" and 500 iron wedges from a John Mounty of Georgetown.[4] Considering the massive amounts of stone needed to be quarried and worked, this was a necessary and ongoing expenditure.

The following tools were commonly used:

1. **Sledgehammers** were usually held by one man with both hands. The tool weighed ten to twenty-five pounds and needed considerable force for driving drills and wedges.

Quarrying Aquia Stone

Tim Buehner, historical architect, drew this picture to show how this area on Government Island would have been quarried in the late 1700s and early 1800s. *National Park Service*

2. **Drills**, also called jumpers, were three-foot-long tools with chisel-like points and were used to cut holes into the rock or divide large stones. One man would hold the drill while another would strike the drill with a sledgehammer. After a blow, the drill would be rotated and the process would be repeated. Hand drills, eight to fifteen inches long, also were used.

3. **Wedges** were usually made of iron or steel and in various shapes. Rectangular ones were used in fissures, or cracks, while a wedge called a plug and feathers was used in round holes.

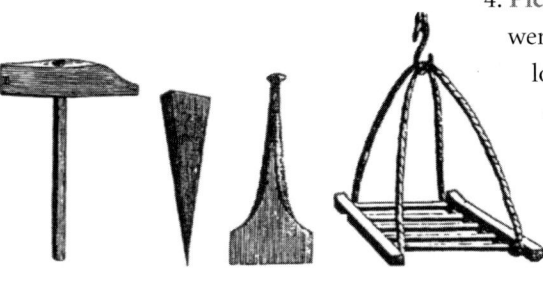

A roughing-out tool, wedge, chisel, and cradle hoist were some of the tools used to remove the heavy Aquia stone from the quarries.
Denis Diderot, Library of Congress

4. **Picks** looked much like pickaxes and were fifteen to twenty-four inches long. They were used for rough dressing at the quarry site. One man would use both hands to swing a pick.[5]

After the stone was quarried, it needed to be moved. Since one cubic foot of Aquia freestone weighs 120 pounds,[6] the workers had a formidable task. The large blocks of stone were removed with wooden derricks or cranes equipped with wooden pulleys and hemp rope. Holes in some existing stone indicate that possibly the legs of such equipment were placed in the holes for support. The stone could then be removed to a sawing pit for further work or dressed according to instructions from Washington, D.C.

Throughout the first ten years of the initial construction of both the President's House and the Capitol, stone was quarried by traditional means. In 1793, the Commissioners received a letter from a Samuel Millikin of Philadelphia who had a "marble and stone machine." He claimed that his "machine will pay its cost in one year - and will do as much as 4 to 5 men…."[7] Hoping that this would help solve the problem of not receiving adequate amounts of stone, the Commissioners asked Thomas Jefferson to look into such a machine and see if it would be feasible for their use.[8] Later, Jefferson replied that he examined Millikin's designs but said that "…I confess it does not appear promising to me. It is certainly inferior to that used in Europe of which there is a drawing in the *Encyclopedie Methodique*."[9]

According to Columbia University Professor Harley J. McKee, mechanical means of quarrying stone were not used in America until the l880s, when steam-powered channeling, or trenching, machines were introduced to quarries in New England.[10] The 200-year-old pick marks upon the quarried rock faces of Government Island, as well as those at nearby quarries, indicate that machinery was never used. However, another quarry a couple of miles down Aquia Creek used machines at the turn of the century. According to Wilbur Segar, whose father was the last foreman of the George Washington Stone Company, machines were used until its closing in 1931.[11] (This company had no connection with the president.)

Quarrying Aquia Stone

Two-hundred-year-old pick marks are still visible upon rock faces.
Lou Cordero, Fredericksburg Free Lance-Star

9

Transporting the Stone

The way stone was transported on or from the government-owned island is without documentation. By looking at the topography of the island today and examining the remnants of facilities, one can speculate how the stone was moved. A stone wharf is still visible on the island's northeast corner, so it is possible that either scows or shallow boats of some sort collected the stone from the wharf for its passage to the district. Leading down to the wharf from the quarry sites that dot the island is a wide path. Since stone was usually placed on skids and pulled by oxen or horse, it appears as if this path, or skid run, was created in such a manner. Other names for skids were sleds, sledges, stone boats, or drags. Gravity was an important factor in the operation of the quarries, for the stone was quarried at a raised elevation and moved to water level.

A short distance downstream from the island is a freestone wall that drops into Aquia Creek. It can be seen only at low tide if cruising past on a boat or looking down into the water from the shoreline. The pick marks are still visible despite the fact that the stones have been submerged for over two centuries. Old land survey maps name the wall as "Stuart's Wharf" or "Stewart's Wharf."[1] This suggests the wharf may have been built or possessed by Robert Steuart, who owned the one acre upon the island.

Steuart's Wharf is located right by Austin Run, a small tributary that flows into Aquia Creek. Milton Dickerson, a Stafford

A skid run, still visible on the island, shows where stone was dragged from a quarry site to the water's edge for shipment to the district.
Author

County resident who sailed on the creek as a boy, said the depth of the water at that juncture used to be twenty-two feet and believed that the submerged stone wall is at least that tall. Dickerson said his father, Captain Benjamin A. Dickerson, used to pilot boats up and down the creek during the early 1900s. Large vessels would go up to that area, known also as Coal Landing, and load or unload their boats, then turn around. According to Dickerson, the large vessels would never go up as far as the island.[2] It can be supposed that in the 1790s the same procedure was used. Stone was loaded on scows, or flat bottom boats, at the island's stone wharf and floated down to the nearby stone wall, or Steuart's Wharf. From there the quarried stone was transferred to larger ships docked in the deep water. Once loaded, they made the trip down the creek and up the Potomac to the district.

This theory is supported in a letter written in 1824 that mentions a wharf east of the island. Thomas Towson had contracted with the government to quarry stone for the column shafts needed on the East Front of the Capitol. He wrote to Joseph Elgar, Commissioner of Public Buildings:

> Sir: I have this evening, taken your fourth column shaft out of the canal, which is now laying on the Schow, the three first I have bond'd on a wharf East of the Island, where the vessel can take them off at any time, she arrives.[3]

Transporting the Stone

The canal to which Towson refers may have been the same canal mentioned thirty-three years earlier, in 1791, when L'Enfant purchased the island. Concurrently with the island's purchase, L'Enfant rented another quarry from a John Gibson. Commissioners' records indicate that a canal was to be constructed connecting the two quarrying facilities, but records, maps, and deeds do not indicate where Gibson's quarry was located.[4]

A canal was also mentioned in 1806 when Benjamin Henry Latrobe, Architect of the Capitol, visited the island. Evidently, the island quarry had been closed for a few years. He stated that if he could find "fine" stone on the island and if he could obtain sufficient funds from Congress to reopen the quarry, stone would be cheaper than that obtained by private individuals. He presented his plan for a more efficient removal of the stone:

> Between the great mass of rock on the island of Acquia and the deep water of the creek is a soft marsh. Through this marsh a canal has been formerly cut, now much choked up, which is barely sufficient to convey the stone by means of a scow to the vessels which bring it up to the city. From the quarry to the canal the stone must be carted. Therefore be thrice loaded and twice unloaded. If the public work were sufficiently certain, it would be highly advisable and very easy to erect a railroad across the marsh where it is narrow, to a part of the creek below the

This aerial view, looking down Aquia Creek to the Potomac River, shows Government Island in the foreground. Vegetation hides the various quarry sites upon the island. The journey down Aquia Creek to the Potomac is approximately six miles. The distance to the district is another forty miles.
Jack E. Boucher, Historic American Buildings Survey

Ruts created by oxen pulling heavy stone along a river bed can still be seen today near Robertson's quarry site. Barbara S. Kirby

island. The rail wagons would take up the stone at the quarry, and unload it in the vessel. But this means…the stone could be procured, even including the expense of the rail way, at least 25 per cent cheaper than at present.[5]

This railroad never came to fruition, but the island was reopened and evidently the canal, too.

George Gordon, Stafford County's former Commissioner of Revenue, said that around thirty-five years ago, an elderly gentleman claimed there was a stone road going from Steuart's side of the island to Stewart's Wall or Wharf. It was under many feet of vegetation then and perhaps is now even under more vegetation.[6] Is the stone road part of the canal system that joined Gibson's quarry to the island? These questions may never be answered because many records were lost or destroyed in the Stafford County courthouse during the Civil War.

Aerial photographs today do not disclose any traces of a canal or road. The years of vegetation growth and dredging around a portion of the island may account for the absence of any visible signs. Therefore, all one can do is speculate as to its hidden location.

Most of the other quarries were located on Aquia Creek, making accessible stone transportation via water. However, one owned by a William Robertson was located a couple of miles inland. Nearby his land was a stream, but it was not deep or wide enough to support a scow. Today sets of wheel ruts can be seen in the stone stream bed that leads to Aquia Creek. Evidently, his quarried stone was placed on carts that were pulled by oxen. Their repeated journeys, carrying such heavy loads, carved the grooves in the stone trail.

The stone that got to the water's edge was taken away by large vessels. The journey from Government Island to the district was approximately forty-six miles, six miles down Aquia Creek and forty miles to Washington. The names of some of the ships can be found in Commissioners' records. For example, a Mr. James Smith was selected by the Commissioners to transport stone from Aquia Creek to the district in May of 1792.[7] In 1795 the Commissioners' records indicated that he was being paid approximately $1.11 per ton and was hauling the stone using three schooners. The amount each could carry was indicated in the records:

> *The Ark* 27 tons
> *The Columbia* 39 tons
> *The Sincerity* 40 tons[8]

Records from January 1796 mention that two schooners, the *Quarrier* and the *Contract*, as well as the sloop *Peggy* brought up 142 tons of stone from Aquia.[9] It is unlikely that these large vessels could carry such great loads and be able to dock near the shallow waters surrounding the island. They needed a deep water harbor such as the one by Steuart's Wharf.

Written correspondence to and from the Commissioners, pay records, and contracts give us the names of men who were in charge of transporting the stone on boats. William Smith informed the Commissioners in 1794 that he did not like being paid by the load for hauling stone up from Aquia; instead, he wanted to be paid $4 a ton. The Commissioners finally relented.[10]

Once the vessels completed their voyage from Aquia to Washington, D.C., they landed at one of two district wharves. One was located below the President's House on Goose Creek. Later called Tiber Creek, it flowed into the Potomac a short distance down from Rock Creek. The contract for the building of this wharf reveals the magnitude of the construction:

> …build a wharf…to extend ninety feet out into the river and in breadth fifty feet, to be raised three feet on a level above the common high tides, the loggs used in building the wharff to be well fastened with two substantial iron bolts in each logg

and where necessary with Trunnels [a wooden peg] also; the wharff to be well filled in with dirt and finished in a compleat and workmanlike manner….[11]

The other wharf was located on the Eastern Branch below the Capitol. Today the Eastern Branch is referred to as the Anacostia River. Quarried stone was the major item delivered to the wharves but not the only construction material. Such supplies as wood for fuel to burn the bricks, lime to create mortar, timber, and nails were also received.

After the stone reached the wharves, its next movement was equally daunting. Bob Arnebeck states in his book about the building of Washington, *Through a Fiery Trial*, "It took six slaves two days to unload one 'shallop load.' And then it took eight slaves three days to put that load on the drags."[12] This method of unloading a ship and putting the stone on the wharf, only to be loaded again on the drags, or wooden sleds, was thought by William O'Neale, a quarry manager and resident of the district, to be too time-consuming and not at all efficient. In 1794, he suggested that a wharf or dock be made where the stone could be placed immediately on the drags and then pulled by horses or oxen rather than be stored on the wharf.[13] Two years after O'Neale's suggestion, stone was still clogging up wharves. Some suspected that it was just a tangible way for members of Congress to see that work on the two structures was proceeding nicely and that stone was available.

John Zweifel of Orlando, Florida, and his family made a fourteen-by-eighteen-foot diorama of the "Building of the White House." This section shows how oxen dragged stone on wooden sleds around the district.
Author

Transporting the Stone

This balloon view of Washington, D.C., shows the Washington Canal, which joined the Potomac River to the Anacostia River, then known as the Eastern Branch. The picture appeared in *Harper's Weekly* on July 27, 1861.
The Kiplinger Washington Collection

A canal for the capital city was originally proposed by L'Enfant that would have aided transportation of the stone, but its construction was postponed, as money was needed for the two federal buildings. After construction of the buildings began, canal diggers were employed, but the digging was not completed. In 1795 there were complaints that stagnated water in unfinished sections created a health hazard. Besides that, the sides were falling in.[14] A year later, people complained about the mess canal diggers had made at the foot of the Capitol. Construction on the canal began in earnest in 1810 by Benjamin Henry Latrobe. He wrote, "We are going this Summer to cut a Canal from the Potowmac thro' the heart of our city to the Harbor or Eastern Branch." It was completed by 1815. At that time, the *National Intelligencer* reported:

> Marble, stone, etc., are now landed at the foot of the Capitol, which otherwise must have been hauled at a great expense from four times the distance. The citizens can now land everything near their doors with considerable reduction of expense, trouble, and time.[15]

The canal paralleled the Mall on the north side, flowing approximately where Constitution Avenue is today. It crossed at the foot of the Capitol's west side and joined with the Eastern Branch or Anacostia River. Stone was unloaded for the Capitol near the site of the present day's Botanical Gardens. The canal needed frequent dredging and was not profitable for the canal company. It was considered an eyesore and filled in during the early 1870s.

10

Work Continues

On April 10, 1792, the Commissioners contracted with William Wright to be the quarry superintendent of the government-owned island. Nothing much is known about Mr. Wright except that he was from Alexandria, Virginia. His job description gives insight as to his many tasks:

> …employ a number of good hands, not exceeding twenty, and oversee and work them diligently in clearing the Stone quaries and [*indistinct*] and loading on board vessels such stone as may be from time to time directed. He is also to make gauges, measure and assist in roughing the stone. He is to be paid a dollar and a quarter of a dollar for every working day of his services. The wages of the men, as well as his own, are to be paid on a pay role to be made out by him monthly and the men are to be found provisions by the Commissioners….[1]

Eight months later, in December, Collen Williamson, master stone mason whose name was on both cornerstone plaques, said that there was plenty of stone that had been quarried, but it had been improperly done. The Commissioners said that it was Wright's fault for not hiring slaves to keep the costs down. Wright wrote back to the Commissioners, "Has there not always been stone ready?"[2] To improve matters, the Commissioners ordered that twenty-five slaves be hired for 15 pounds a year. Then on January 1, 1793, they fired Wright and put Williamson, who was in his sixties, in charge of the island quarries.[3]

Birthstone of the White House and Capitol

Up in the district, work was progressing at both the President's House and the Capitol construction sites. By autumn, Hoban was already looking ahead to the winter and the spring of the following year. He recommended that skilled stone workers from the district move to the quarries during the winter months in order to be constantly employed. He was fearful that if they did not have a job during the cold winter months, they would attempt to get work elsewhere and the Commissioners would not be able to hire them back when the weather improved.[4] Instead of moving the skilled workers to the quarries as Hoban had originally suggested, Williamson was asked to have the men work in the sheds and dress stone for use in the spring. Since work was reduced in the winter, the men's wages were decreased, too.[5]

Also in 1793, Hoban informed the Commissioners that 1,350 tons of freestone had come from Aquia in thirty-eight loads. He was concerned that there would not be enough stone for the spring of 1794 since "a large proportion" of the stone was of poor quality and the cost of the stone was too high.[6] Since good freestone seemed difficult to obtain, the Commissioners decided in August that the walls of the buildings would be constructed of brick and faced with freestone rather than be made entirely out of Aquia stone.[7] The walls would have a combined thickness of approximately four feet. This decision by the Commissioners greatly upset Collen Williamson, who wanted to continue constructing the Presidential Palace entirely of stone.

> *In August of 1794, it was reported that the first story of the President's House was completed.*

Williamson was no longer in charge of the quarries. Taking his place was a young Englishman by the name of George Blagden. Blagden had first worked at the President's House in 1793 and was later moved over to the Capitol. He had the distinction of being one of the few men who labored on friendly terms with all the warring factions. While architects and Commissioners, Presidents, and Congressmen fought over costs and designs, he went to work. (For thirty years he worked at the Capitol. Unfortunately, he died by accident when a bank of earth at the south angle of the Capitol caved in upon him.[8] A Commissioner wrote Charles Bulfinch, who was then Architect of the Capitol, and said, "We have met with an irreparable loss."[9])

In August of 1794, it was reported that the first story of the President's House was completed. It consisted of "all stone in and out, four feet below ground and 12 feet above it." After this the other stories would be constructed, as ordered by the Commissioners a year before, with brick and freestone.

During 1794 a need emerged for skilled masons. The Commissioners asked John Greenleaf, a speculator, to get European workers who "…have been bred to cutting and laying free stone." They were to work for two to four years and would get food, board, and passage. Workers with families would not get food but could use land located northeast of Massachusetts Avenue for gardens. Single men would receive not more than 30 guineas a year and married men 55 guineas.[10]

The year of 1794 was not a good one for Collen Williamson. He was already having difficulties with Hoban, but troubles magnified with the appearance of Cornelius McDermott Roe, a stone contractor. First, Roe did not complete his apprenticeship as a mason, and many masons refused to work for him, for they considered him unqualified. Second, Roe proposed to do stonework at a fixed price by the amount of stone actually laid, on a piecework basis. He was hired by the Commissioners to do just that at the Capitol. Therefore, Williamson and his men, who were working for wages, returned to the President's House.

When Williamson's masons discovered that Roe's masons were getting more money for piecework than they were for equivalent work, they wanted a raise. Williamson, upset by Roe's interference with the masons, brought grievances before the Commissioners. The Commissioners did not increase wages, but to pacify Williamson, gave him supervision over Roe's masons. Hoban, an ally of Roe, told the men who would not do piecework to leave, and they did. Furthermore, Roe's men would not listen to Williamson. This infuriated Williamson, who claimed that Hoban and Commissioner Carroll, both Catholic, were forcing others to leave to make way for "a Passel of Irish Papists."[11]

The Commissioners paid no attention to a complaint from the masons that said, "McDermott Roe's [work] is totally unfit for such a building, and must be undone or the House [Capitol] will be ruined."[12] By 1795, Williamson's predicament worsened as problems increased. European masons were complaining. They were not used to working and training slaves, and Roe increased the number of slaves to fourteen; masons numbered eighteen. Hoban kept saying that there was not enough stone and that it was Williamson's fault. All this caused Williamson's termination. Hoban was put in charge of all masons in the city.[13] Later, to no avail, Collen Williamson asked for his old job.

Prior to 1795, the Commissioners hired individuals to quarry or ship stone. Once stone arrived at one of the Commissioners' wharves, slaves would be taken from building sites to move the freestone to the required locations. Stone was moved on scows and carts owned by the Commissioners. This costly process took needed laborers away from necessary tasks. Therefore, in 1795, the Commissioners changed their method of operation. Now they dealt with contractors or individuals who were not only required to quarry the stone but had to ship it as well as deliver it to either building site. The Commissioners even sold their scows and carts. Two contracts were made for freestone with Cooke and Brent from Aquia Creek and another with a quarrier from nearby Chopawamsic Creek. The Commissioners were very pleased with the contracts, for they expected 5,000 tons in 1795 and 6,500 tons in 1796, quarried, shipped and delivered for "at least 7/6 [$20] per ton less than it has hitherto cost us."[14] Unfortunately, for the Commissioners, the stone from the Chopawamsic area proved to be unsuitable for use in the district.

The year 1795 was extraordinarily difficult due to accidents and delays. For example, during the summer when masons were laying a freestone wall on the foundation of the Capitol's north and south wings, it collapsed. Some of the bad foundation walls had to be taken down and rebuilt using "large bond stone." Most of the season's work had to be redone at a cost $1,264 for the north wing and $1,470 for the south wing.[15] There was talk that it was the fault of McDermott Roe and his men.

In 1795 the Commissioners hired George Hadfield as Superintendent of Construction at the Capitol. They were hesitant at first since he was only thirty-one years of age. Hadfield, an English subject, was born in Italy. His schooling was in both England and Rome. The famous American artist John Trumbull, a friend of Hadfield, recommended he be hired. Once the young architect arrived, he began to criticize the work already completed on the Capitol as well as Dr. Thornton's designs.[16]

Another delay occurred in the fall of 1795. George Blagden checked some of the stone that had arrived in the city. It had come from Aquia Creek quarry sites on the island. Blagden discovered that around thirty tons of it was either too coarse or too soft and therefore unacceptable.[17]

In October of 1795 it was found that the quarries were sending ashlar, or squared stone suitable for facing the buildings, but not bill stone, which was fine-grained and cut for a specific area such as a cornice or frieze. Blagden visited the Aquia quarries and reported, "'Twas a little mortifying to be informed that two of the vessels were just loaded with that kind we so much abound in. And it was by mistake."[18]

By the end of the summer of 1796, construction of the two structures seemed to be proceeding with fewer interruptions. President Washington was given a party when he visited the city in August and informed that the Capitol walls were going up rapidly due to the fact that fifty-one white laborers and about as many slaves were now aiding the masons.[19]

In the fall of 1796 the Commissioners wrote two letters to Cooke and Brent of Aquia. Stone was needed if work was to continue. They said that the stone ordered six months before had not yet been delivered and if it was not delivered in eight days, they would be held responsible.[20]

Building continued until labor unrest arose during the spring of 1797. Carpenters who were constructing the President's House roof and five stone carvers asked for a raise. The stone carvers were putting up the decorative capitals along the 350 feet of architraves. The carpenters did not receive a raise, but the stone carvers got an increase of 17 cents a day. Their daily wages were now $1.63.[21]

In 1797 Collen Williamson, fired two years previously, wrote a long letter to President John Adams. Still upset by the way the buildings were being constructed, he claimed he could have saved 20,000 pounds [money] by using stone alone instead of the brick-stone combination:

> …I told the commissioners they wor not aware of the expense that bricks would lead them into, that I had considered the Difference…. Besides bricks are never built in such works where stone is to be got, the bricks have Nither the solidity nor durabilety of stone, and likewise one half of the mortar will lay the stone, that the bricks, will require….[22]

He also wrote Adams that the Commissioners were unfit "although that great and good man appointed them." He wrote that appointing inexperienced Hoban was like "the blind leading the blind…and…if they wanted boots or shoes they would not apply to an Irish carpenter to make them…."[23]

In February 1798 an accident delayed work on the balustrade and architraves at the Capitol. A forty-ton load of stone departing the island sank in Aquia Creek. The stone was not only loaded for the Capitol, but also for the completion of Commissioner Daniel Carroll's residence.[24] A little later that month, the Commissioners wrote to Cooke and Brent requesting more stone. "Our hands are entirely stopped for want of Stone, we entreat you to send a Cargo immediately."[25]

This untimely accident compounded problems for the Commissioners. They were already low on funds to complete their building projects for the 1800 deadline. Therefore, they decided not to advance the wages of workmen during the traditionally high construction period of the first of March through the first of November. This

> *In the fall of 1796 the Commissioners wrote two letters to Cooke and Brent of Aquia. Stone was needed if work was to continue.*

meant significant reduction in the workers' earning power and created a major labor dispute.

The stone carvers at the Capitol wrote to the Commissioners and asked them to reconsider their decision:

> Finding Ourselves agrieved by our Wages being Curtailed, We take this Method to lay our Case before ye and Expects after your further Consideration on the Business that ye will remove our Grievance by Raising our Wages to the Same as it was last Summer…we Expect Your Answer will be Conformable to our wishes….[26]

That day the Commissioners also received a letter more threatening from the Capitol stone cutters. They resolved "…to Quit the work At the lattest of this Month unless you agree to Raise our wages to the Same it was last summer and to Continue unto the first of Novembr."[27]

The next day, April 17, 1798, the stone cutters at the President's House also sent a letter expressing the fact that they considered the actions of the Commissioners quite unfair.

The response of the Commissioners to the stone cutters was immediate:

> Gentlemen
>
> We have your application on this day before us and as we have no intention of raising your wages; we take the earliest opportunity of letting you know, that we shall expect you to quit the public employment at the end of the present month and all the buildings you now occupy, belonging to the public, as they will be wanted for other Stone-Cutters; whom we shall take the necessary measures to engage. You will therefore quit the Buildings by the last of the present month.[28]

It is significant that the stone cutters were fired but not the stone carvers. Apparently, the Commissioners were informed that stone cutters from Philadelphia could be hired for less.

Requests for more stone cutters were sent to Philadelphia and to Robert Steuart of Baltimore, the same Robert Steuart who owned the one acre upon Government Island.

> Washington, 18th Apl. 1798
>
> Sir
>
> It is not improbable but we shall want some Stone Cutters & Setters at the public Buildings, about the first of May. I will thank you to inform me, by the mail as early as you can, what are the wages now given at Baltimore to journeymen stone Cutters and Setters, and what number can probably be procured at your place.[29]

Work Continues

At the end of the month, George Hadfield, superintendent of construction on the Capitol, proposed a compromise. In his letter to the Commissioners, he stated:

> …Some of the stone cutters at the Capitol have been with me and expressed a wish to come to a reconciliation with the Commissioners; I believe it is the better of the whole. And I promised to inform you of it.[30]

Wages were increased for the peak work season, but they were not as high as those of the previous year. Thus, a compromise solution averted any more delay.

(After two and a half years, Hadfield was dismissed of his duties and Hoban was put in charge of construction for both buildings. Hadfield, however, stayed in Washington and designed many buildings, such as the Treasury and City Hall. Unfortunately, the Treasury was burned by the British in 1814. Despite his success, he never rose above his failure at the Capitol.[31] It was later written of him, "Loiters here, ruined in fortune, temper, and reputation." Hadfield died in the capital city at the age of sixty-two.)[32]

Work on the Capitol continued. Polish visitor Julian Niemcewicz described its expansive appearance in May of 1798:

> A huge scaffolding surrounded it and all around for a considerable distance the ground was covered with huge blocks…of stone, some already cut and polished, others yet undressed. The lower part of this picture was composed of some sheds for cutting the stone and for working on the roof and the framework, some cabins scattered here and there, a shelter for the workers, two or three small shops for liquor and other articles of prime necessity. The top of the edifice was covered with 200 workers, raising the stones by means of machines and placing the first framework of the roof. All were working in silence.[33]

> *After two and a half years, Hadfield was dismissed of his duties and Hoban was put in charge of construction for both buildings.*

Meanwhile, work on the exterior of the President's House was nearing completion. In fall 1798, Hoban reported that the workers were "cleaning down and painting the wall of the building and striking the scaffolds."[34]

11

Moving into the Federal Buildings

Nine years had passed and the ten-year deadline for the completion of the two structures was about to occur. Unfortunately, the guiding force behind the buildings, President Washington, was not able to witness their completion, for he passed away December 14, 1799. However, he did write in his diary on November 9, 1799: "Set out a little after 8 'clock-viewed my building in the Fedl. City-."[1] It is touching to see how he affectionately referred to the Presidential Mansion as "my building."

John Adams was the first chief executive to live in the President's House. The house, however, was incomplete when he and his wife moved in. It was merely a shell. His wife, Abigail, wrote to her sister about her new living quarters. Her description paints a stark but realistic picture of the house and its surroundings:

> As I expected to find it a new country, with Houses scatterd over a space of ten miles, and trees & stumps in plenty with, a castle of a House-so I found it-The Presidents House is in a beautifull situation in front of which is the Potomac with a view of Alexandr[i]a. The country around is romantic but a wild, a wilderness at present.
>
> I have been to George Town...It is the very dirtyest Hole I ever saw for a place of any trade, or respectability of inhabitants. It is only one mile from me but a quagmire after every rain. Here we are obliged to send daily for marketting; The

capital is near two miles from us. As to roads we shall make them by the frequent passing before winter, but I am determined to be satisfied and content, to say nothing of inconvenience &c....

The President's House was completed in 1800 and became the largest and grandest house in the new nation. Kollner, White House Historical Association/White House Collection

This day the President meets the two Houses to deliver the speech. There has not been a House untill yesterday-We have had some very cold weather and we feel it keenly. This House is twice as large as our meeting House. I believe the great Hall is as Bigg. I am sure tis twice as long. Cut your coat according to your Cloth. But this House is built for ages to come.... Not one room or chamber is finished of the whole. It is habitable by fires in every part, thirteen of which we are obliged to keep daily, or sleep in wet & damp places.[2]

Dr. William Thornton, designer of the Capitol, and his wife went to view the newly completed President's house on March 20, 1800. An entry in Mrs. Thornton's diary gives a description of the grounds. "...it is at present in great confusion, having on it old brick kilns, pits to contain water used by brick makers, rubbish &c&c."[3]

The grounds surrounding the Capitol probably looked much the same, for by the winter of 1800 only the North Wing was finished and ready for the Congressmen. The new building appeared to be a boxlike structure with dimensions of 126 feet by 121 feet, 6 inches.[4] Its lines, however, were simple and dignified, an example of Italian Renaissance architecture. Three sides were built of Aquia stone while the south side, facing the soon-to-be built Rotunda, was temporarily constructed of brick. Inside, the gentlemen were greeted with wooden partitions and wooden flooring. The structure was topped with a wooden roof. This new two-story building with basement would house the Senate, the House of Representatives, the Supreme Court, and the Library of Congress!

By 1800 this boxlike wing of the Capitol was nearly completed. Notice the masons cutting Aquia stone in the foreground. This structure was cramped, as it held the Senate, House, Supreme Court, Circuit Court, and Library of Congress.
William I. Birch, Library of Congress

The rest of the Capitol building was under way. Foundations had been laid for the Rotunda. The House Wing's first floor, still under construction (currently Statuary Hall), was just a few feet tall.

Despite its rough and unfinished appearance, John Adams referred to the Capitol as a "temple" in his State of the Union message:

> It would be unbecoming the representatives of this nation to assemble for the first time in this solemn temple without looking up to the Supreme Ruler of the Universe and imploring His blessing.[5]

After a short time, the House of Representatives found that they were just too cramped and ordered construction of a temporary brick building. The new House "Oven," as it was called, was located on the south-wing side. They moved into the Oven in 1801 and met there until it was demolished in 1804. The House then returned to the original north wing for three more years until their new wing was ready for occupancy.[6]

12

Benjamin Henry Latrobe's Contributions

We know much about Aquia freestone, quarries, workers, and buildings in the Federal City because of Benjamin Henry Latrobe. Born in England in 1764, he arrived in America at the age of thirty-two. Trained as an English architect, he was appointed by his friend, Thomas Jefferson, to become Surveyor of Public Buildings in 1803. This required him to work on both the White House and Capitol.

Latrobe kept sporadic journals. His detailed entries give insight to early American life. His words, pen and ink sketches, and watercolor drawings help us visualize the beginning stages of our nation and its flora and fauna.

FREESTONE

In 1806 Latrobe visited outcroppings of Virginia sandstone. He wrote about "freestone, of which the public buildings in the federal city are built…." Some of his findings were:

> The free stone, is found in immense disrupted Masses, which show themselves on the side of Gulleys formed by Watercourses.[1]

> …hard sound Mass, of great depth and thickness, the Rock suddenly becomes a mere friable Land Mass. Quarrying therefore is often a lottery in which the blanks are more numerous than the prizes.

All of it expands when wet, and contracts when dry. This property it seems never to lose, though buried ever so long in the Walls of a building, unless, as in the Capitol it is controuled by the excessive weights of the incumbent Mass.[2]

…if it once becomes dry, and remains sound, it has never been known afterwards to fail.[3]

…nodules of clay, the pebbles, the gravel, the nodules of Iron ore, the lumps of quartz….[4]

OTHER STAFFORD QUARRIES

Cooke and Brent Quarry

In 1806, six years after completion of the President's House and after both Congressional houses resided in the Capitol, Latrobe visited the Aquia quarry of Cooke and Brent. Their quarry appeared to him to be very thin in strata, perhaps because so much was already quarried out. He stated, "Good excellent stone 8 to 2 feet…the stone is exceedingly good, and free from Iron ore and clay holes." He believed Robertson's quarry was superior. "Mr Robertsons quarry is now the best in Wor…Sound rock…15.0 [feet]."[5]

Benjamin Henry Latrobe, 1764–1820 by Charles Wilson Peale.
White House Historical Association/White House Collection

Robertson's Quarry

The Robertson to whom Latrobe referred was William Robertson, who owned a quarry about three miles from Government Island. The exact date he opened his quarry is not known. Presumably, it was after the initial district structures were completed, as Robertson made an 1804 agreement with Latrobe. His quarry was close to Austin Run, a stream that flowed into Aquia Creek. Today a housing development envelopes the quarry site. But still visible are Robertson's foundation and chimney. A trip to a nearby stream discloses trenches created 200 years ago by oxen dragging carts to Aquia Creek.

Robertson's home was visited by Latrobe in August of 1806. He wrote of his experiences spending the night in a little log house and sketched the house, out-buildings, and quarry.

Benjamin Henry Latrobe's Contributions

Latrobe recorded his visit to Robertson's quarry in drawings as well as words.
Maryland Historic Society

Here is his account:

> He [William Robertson] therefore fixed upon a beautiful little knoll in the midst of the woods close to his quarry, and determined to form a settlement there.... A log house kitchen, a do. stable, smithshop, hen house, meat house, and tool house, sprinkled irregularly over the knoll, gave to the whole a some what romantic appearance.[6]

STONE CUTTERS

Even as late as 1806 stone cutters were needed to work at the Capitol. Latrobe hired five stone cutters in Philadelphia and two in Baltimore. He wrote to Thomas Jefferson, "As far as I can ascertain there is now not an unmarried stone cutter left" [in Philadelphia].[7]

Birthstone of the White House and Capitol

Latrobe's journal discussed class differences between workers and educated men like himself:

> I was the only person in the packet whom the American republicans ever call a Gentleman. The other passengers, 10 in number were young mechanics of remarkably decent manners. Two of them were Stone cutters of Philadelphia whom I had engaged for the Capitol, two were from Wilmington, Del.… I was treated by them with very marked respect, and perhaps my presence controuled their spirits and induced them to act with caution. I could not however help observing how very different the language of civility, and indeed the whole phraseology of these citizens is from that which is used in what are called the upper

The foundation of Robertson's house and chimney, sketched in 1806, are still standing today.
Maryland Historic Society

circles of Society. Thus inequality even in the most unessential affairs of mankind grows up in society, in spite of laws and regulations which necessarily mix together all classes of citizens on very numerous occasions, and oppose, as far as regulations can oppose every growing separation of our citizens into orders and ranks."[8]

INHABITANTS OF WASHINGTON, D.C.

A frightful experience in the nation's capital caused Latrobe to contemplate the fate of its citizens. One day in 1806, a little before sunset, Latrobe was walking to the home of Robert Brent, the mayor of Washington, D.C., and relative of the Aquia Brents. "…As I passed over the uninhabited part of the town between the Capitol and his house, which is a low swampy piece of ground covered with Bushes, a tall middle aged woman popped out upon me from a cross road with a Gun in her hand." She turned out to be a widow with children and wanted Latrobe to buy the gun so she would have money to feed her family. Latrobe continued, "…But what is to become of a widow with sick children in this wretched and desolate place, when the present temporary relief is expended! The City abounds in similar cases. The families of Workmen, whom the unhealthiness of the city and idleness (arising from the capricious manner in which the appropriations for the erection of the public buildings have been granted, giving to them for a short time high wages, and again perhaps for a whole season not affording them a weeks work) have ruined in circumstances and health, are to be found in extreme indigence scattered in wretched huts over the Waste which the law calls the American Metropolis, or inhabiting the half finished houses, now tumbling to ruins which the madness of speculation has erected." Latrobe continued and talked about "Master Tradesmen, chiefly building artisans," who purchased lots, built houses, and invested their all only to find that "Distress and want of employ has made many of them sots…."[9]

13

Other Freestone Quarries

The contractors for the federal buildings relied upon several main suppliers of freestone. Many quarriers desired government contracts, but few were awarded them because much of the freestone offered was inferior.

THE GOVERNMENT'S QUARRIES

During the initial construction period of 1792–1800, the quarry sites on Government Island contributed most of the stone, especially for the White House. When L'Enfant purchased the island in 1791 from the Brents, he also arranged for a ten-year rental of a nearby quarry owned by John Gibson. Both quarries were usually referred to as the "Public Quarries." The rental agreement with Gibson was terminated after six years. The Commissioners wrote, "…we are nearly done with [your] free stone" and wanted him to "Take the quarries off our hands."[1] After that, the term "Public Quarry" or "Island Quarry" was used interchangeably with "Public Quarries."

COOKE AND BRENT

Once the Brents sold the island to the federal government in 1791, Daniel Carroll Brent opened another quarry operation with John Cooke. They became private contractors doing business under the name of Cooke and Brent. Historians have not determined the quarry's exact location, but it seems to have been close to the island with access to Aquia Creek.

Cooke and Brent were an illustrious duo. Both were militia colonels. Cooke was married to Mary Thomson Mason, daughter of George Mason IV, author of the Virginia Declaration of Rights, from which the United States derived its Bill of Rights. Brent was from Stafford County and a member of the Virginia House of Delegates.

They also managed the public island. A *Virginia Herald* advertisement, printed in Fredericksburg and dated December 22, 1794, stated:

> Wanted to Hire, for the next year, to work on the FREE-STONE QUARRIES, lately occupied by the Public, on Aquia Creek, Sixty strong, active NEGRO MEN, for whom good wages will be given - They shall be well used and well fed.
>
> Daniel C. Brent
> John Cooke
> Stafford, Co.[2]

> *"The public Quarry, is opened and promises well.... I flatter myself to be entirely supplied from thence."*

Cooke and Brent's own quarry seemed to stay quite active from the 1790s through 1804. Records indicate, however, that the island was not used much after 1800. Cooke and Brent indirectly forced Latrobe to reopen the island quarry in January of 1804, as they requested that the government pay $10 a ton for quarried stone. Since this was $2.34 more than they had received the previous year, Latrobe said he would reopen the public quarry rather than purchase stone at their inflated prices.[3]

In February, Latrobe wrote, "The public Quarry, is opened and promises well…. I flatter myself to be entirely supplied from thence." The quarry did remain open but was not the complete supplier of freestone.

Latrobe offered Cooke and Brent $8.10 rather than the $10 they requested. The offer was accepted. Despite their demands for a pay increase, Latrobe apparently trusted the team. He wrote, "These Gentlemen are the only contractors who may with certainty be relied upon."[4]

PRESIDENT WASHINGTON'S QUARRY

In December 1793, President George Washington received a freestone sample and a letter from Commissioner Daniel Carroll. The sample was from a quarry located on the President's own Mount Vernon property. It appears the Commissioner was inquiring whether Washington would allow use of his stone in the Federal City. Washington said

Other Freestone Quarries

that he always knew "that the River banks from my Spring house, to the Ferry…, were almost an entire bed of free stone; but I had conceived before the late sample came to hand, that it was a **very soft** nature." He requested "an investigation of the Banks by skilful, and orderly people." The President added that he had no objections to open it up as a public quarry.[5]

William O'Neale, working in Pennsylvania, was chosen to inspect Washington's quarry. Records indicate that he was to be paid $60 a month to develop Washington's quarry as well as one in the city.[6] (The city quarry was probably that quarry located around Key of Keys, which contributed foundation stone or bluestone.)

From Washington's writings it appears that O'Neale never came. The President wrote letters in July, September,[7] and December of 1794 asking questions such as, "How does Mr. [Oneil] like the appearance of the Quarry…?"[8] "What has he raised and what has become of it."[9]

In October of 1796, Washington asked a William Pearce to have earth removed from the stone quarry so that George Blagden could examine it.[10] William Pearce responded about three weeks later by writing, "I have had Enough of the Stone made Bare for the person to examine It But as I Mentioned In my last letter I fear the Body of Earth that Covers It is too Great to Work It to advantage."[11]

Apparently the stone was too soft to use or under too much earth to quarry economically. The freestone from the President's quarry was probably never used in the Federal City.

JOHN DUNBAR'S QUARRY

One Aquia Creek quarry that apparently did not fare too well was John Dunbar's. His quarry was directly across the creek from the government's island. Mr. Dunbar had contracted with the Commissioners in 1794 to send stone to the district. However, several months into the agreement, he wrote letters to the Commissioners mentioning he was having difficulty delivering stone. He requested $1,000 in advance so he could fulfill his contract.[12]

While Dunbar was writing to the Commissioners, he was also holding a public auction. In the *Impartial Observer and Washington Advertiser,* he notified the public he was auctioning "80-85 tons of Freestone."[13]

In 1796 Mr. Dunbar, in the Dumfries, Virginia, jail, wrote two letters to the Commissioners giving the reasons for his confinement and requesting payment.[14] That year, a John Henry & Company must have taken over Dunbar's quarry, for Henry advertised in Georgetown's *The Columbian Chronicle* that he had a "freestone - stone cutting business at John Dunbar's Quarry on Aquia Run…." A month later he placed another ad in the same newspaper in which he boasted to have freestone "Warranted **equal**, if not **superior** in quality and colour to any on the continent." The advertisement added that the company could provide "ornamental parts" for houses, "tomb and head stones, chimney pieces, steps, platforms, etc…" to any American seaport.[15]

ROCK RIMMON

Located on a bluff, about two miles southeast from Government Island, was a quarry named Rock Rimmon. Its name probably derived from the rock formation mentioned periodically in the Old Testament. Records indicate that it also went by the name of Rock Raymond.

Thomas Barrett Conway quarried the area in the 1800s. He negotiated with George Blagden in 1804 to deliver 300 tons of stone for $8 a ton. It was to be delivered to the Eastern Branch and was intended for use in the construction of the "South wing" of the Capitol. Latrobe wrote of Conway, "[He] has good stone, but having no great force, may not perhaps compleat his Contract…."[16] (See chapter "Government Island and Other Quarries Close" concerning changes at Rock Rimmon.)

STEUART'S ACRE QUARRY

When Pierre L'Enfant purchased Brent's Island, he purchased all but one acre. That acre belonged to Robert Steuart. Latrobe wrote that Steuart was the "proprietor of an acre, containing the best Stone on the Island." His son, Colonel William Steuart (1780–1839), apparently inherited the acre and quarried its stone for the Capitol in 1804 and 1805. Some of it, however, was rejected, as it fell to pieces upon drying. Latrobe described Steuart's son as "the most wealthy and respectable Stone cutter at Baltimore." The son supplied stone for Latrobe when he was constructing the Baltimore Exchange and Washington Canal in 1816. He also supplied stone for a burial monument in New Orleans that Latrobe was designing for the family of Governor Claiborne. The younger Steuart became mayor of Baltimore and also served in the Maryland House of Delegates.[17]

ROBERTSON'S QUARRY

William Robertson's quarry was located inland, several miles from Aquia Creek. Latrobe wrote during his 1806 visit that Robertson had signed a government contract in 1804. As this was after completion of both the President's House and the Capitol's North Wing, it is unlikely that Robertson's quarry contributed stone until after 1804.

Robertson most likely had quality stone. Latrobe stated in 1804 that Robertson had contracted to deliver "…extra-fine stone for the Cornice and Capitals at $9." He might have had exceptional stone, but Latrobe also wrote, "Robertson is not entirely to be depended upon."[18] (See chapter "Benjamin Henry Latrobe's Contributions" to read more about Robertson's Quarry.)

EDRINGTON & MONCURE

Located close to Rock Rimmon was the quarry of John Catesby Edrington and William A. Moncure. From family papers and quarry records located at the Virginia Historical Society Library in Richmond, it appears that their quarry operated from 1836 to 1839. Their account books list such things as the tools, quarry hands, food, and needed supplies but do not state where their stone was shipped. Perhaps Edrington and Moncure just quarried stone for the local community, but they may have contributed stone for the Old Patent Building and the Treasury in the District of Columbia, as these buildings were being constructed in the 1830s.

Evidently, John Catesby Edrington contributed stone for the Capitol at an earlier period, as his name appeared as a contractor alongside the names of two gentlemen. The contracting teams of "Perley and Edrington" and "Edrington and Stone" quarried stone for the Capitol.

OTHER QUARRIES

Newspaper advertisements provide the names of other quarries lying along Aquia Creek. For example, the *Washington Gazette* ran an advertisement in 1797, "Free Stone Quarry to be rented on Aquia Creek, belonging to Est of Charles Porter, dec'd, Known as Millar's Quarry."[19] The newspaper *Centinel of Liberty* ran an advertisement in March of 1799, stating, "To be sold at The Union Free Stone Quarry on Aquia Creek, Stafford Co, Virginia A Large Scow bottomed built Schooner."[20]

> *Along the Rappahannock River were other sandstone quarries with stone used in many structures in Fredericksburg and in southern Virginia.*

Along the Rappahannock River were other sandstone quarries with stone used in many structures in Fredericksburg and in southern Virginia. The Rappahannock freestone was not as white in color as Aquia stone, but some of the owners tried to contract with the government anyway. One such person was George Richardson. He claimed to "have enough [stone] to build the City of London." George Blagden examined Richardson's quarry in 1804 and reported to Latrobe. Later Latrobe wrote that Richardson was "something of a *braggart*. For how could he assert that he had Stone sufficient to build a city, if it is clear only 4 feet in depth. I hope he may be better depended upon what he *does* than in what he says." On February 17, 1804, Latrobe notified Richardson that the government would not be using his stone for the Capitol.[21]

CONTRACTORS FOR FREESTONE USED AT CAPITOL

A list of suppliers of freestone for the Capitol, from the Office of the Architect of the Capitol, renders no discernible connections except geographic proximity to Aquia Creek in Stafford County. Numbers and notes illustrate connections.[22]

1. George Waller (Intermarried with #6 and #20. A relative, Withers Waller, transported Aquia stone to the district in the 1820s.)

2. James Morton (Married Lucy Horton of Horton's Landing, #6. Landing northeast of Government Island.)

3. James Hewitt (Listed in 1820 census. Did not own land. Perhaps a subcontractor or involved in stone transportation.)

4. Joseph M. Johnson (Not a major participant)

5. Peyton and Dent (Peyton intermarried with #11—Peyton and Cooke. Dent's Landing was on Aquia Creek, across from Government Island.)

6. Hugh Adie (Intermarried with #1. Sheriff and large property owner. Relative Charles Adie was a stone cutter.)

7. John H. Suttle Co. (Intermarried with #23—an Agnes Suttle married a son of Thomas Towson. A Suttle was mentioned in Latrobe's writings—August 22, 1806—as manager of Robertson's quarry.)

8. Edrington and Stone (John Catesby Edrington married Elizabeth Hawkins Stone. James W. Stone, same as #19.)

9. William Stewart (Mentioned previously in Chapter 13)

10. John Carter (Early settler in Stafford; owned considerable property. A joiner by trade.)

11. George Cooke (Mentioned in 1820 census. Son of John Cooke of Cooke and Brent.)

12. William H. Tyler (Owned small quarry near island. Involved with the building of Stafford's jail.)

13. Cossom Horton (Horton's Landing on Aquia Creek.)

14. John Griffis (Owned many lots in Stafford around Aquia Creek area known as Widewater)

15. Perley and Edrington (John C.) (Edrington owned quarry by Rock Rimmon. Called Myrtle Grove in 1830s. See #8)

Other Freestone Quarries

16. John Bramel (Owned small amount of property. He leased land across the creek from Towson #23. Probably dealt in transportation.)

17. Rouzee Peyton and Co. (Peyton family owned vast amounts of land in Stafford near island. See #5)

18. J. Cooke and Co. (James Cooke, see #11)

19. James W. Stone (Ancestor was Maryland Governor William Stone. See #8)

20. John Hore (Intermarried with #1, #6, and 22. Grandson of #1)

21. F. W. Tallifur note: Voucher #7, Sept. e, 1819, Ck. #23 (Notes are from Capitol's records. Minor participant.)

22. John Moncure, Jr., Exe. of Horton note: May 13, 1822, Ck #21 (Preceding notes from Capitol. Moncure was executor of Horton's estate. See #13)

23. Thomas Towson (Stafford land and quarry owner. Owned quarry once owned by John Dunbar and by John Hore, #20. Quarried stone on island for Capitol's columns on East Front in 1820s. Intermarriage with #7.)

24. Isaac Evelith, August 10, 182_ [Last numeral not discernible] (Not major participant.)

25. John W. Baker (Bought Brent family mill in Widewater.)

26. William Lamb (Only Lamb data is from a Stafford tombstone with the name of William Lamb's wife, Mary. Died February 1, 1815, at forty-eight years of age. She was a slave. He worked in a local quarry and allegedly cut and carved her tombstone.)

14

Cutting and Carving Aquia Stone

Most stone reached the building sites from the Aquia quarries in a fairly rough state. It had to be dressed or finished. One way of smoothing a stone was by using a saw. The saw most commonly used was toothless, much like the wire across today's cheese cutter. This toothless blade was strung between the ends of a U-shaped wooden frame. Wet sand was applied between the stone and the blade causing an abrasive action when the saw was drawn repeatedly across the surface. Periodically, the wet sand was replenished.[1]

Plain surfaces could be further smoothed by sprinkling sand upon the stone and then rubbing with a smaller, flat stone. This sanding, for example, produced the smooth surface of those facing stones on the upper floors of the President's House.

Freestone could be cut in any direction by chisel. Chisels were used for delicate work by stone carvers and bolder work by masons. Rough masons struck chisels with a hammer, while skilled stone cutters gently tapped the chisel with a mallet or hammer as it moved along the surface. An accomplished craftsman would not lift the chisel until it traveled completely across the stone.[2] Some chisels were notched and

Birthstone of the White House and Capitol

Left: Craftsmen created by hand both tool-grooved and smooth surfaces as shown here on the ground floor of the White House.
Author

Bottom: Another section of the Zweifel diorama depicts workers laboring on scaffolding.
Author

produced fluted designs. If one examines the exterior ground-level walls of the Presidential Mansion, it appears that the decorative surface was etched using a large-toothed comb. Since this would be impossible, it is mind boggling to consider the time and skill needed to chisel such straight furrows.

Once stones were cut, they were lifted into place. The task of moving large pieces of stones was enormous since they weighed from 300 to 3,000 pounds.[3] If stones could not be lifted into place or hoisted up ramps supported by scaffolding, derricks were used. Two were used at the President's House, each made of wood and fifty feet tall. They were moved wherever needed.[4] Such derricks were also probably used during the Capitol's initial construction. Since the Capitol took decades to complete, masons there used more advanced machinery as time progressed and better equipment was invented.

In order to hoist a stone on a derrick, a method was employed previously used in ancient Greece (ca. 500 B.C.)

Cutting and Carving Aquia Stone

A dovetail cut was made in the top of the stone creating a hole, groove, or slot that was wider at the bottom than the top. Then a tool called a *lewis* was inserted. The lewis consisted of an iron tenon made in sections. Section by section, the lewis was tightly fitted in the groove. Then an iron pin was reinserted through the sections, making the lewis secure. Hoisting or rigging ropes were attached to the lewis. Once the stone was lifted and slipped into position, the lewis sections were removed one by one.[5]

A lewis helped lift heavy blocks of stone.
Tim Buehner, National Park Service

The term "slipped into position" is correct, for the craftsman were so adept at measuring the stone that very little mortar was used between the joints. The fitting was so accurate that the space between the stones was only about one sixteenth inch or less.[6] Considering the massive weight, this was some feat. This accuracy of fit was achieved, in part, by the making of a template. The desired shape was painstakingly drawn on paper and then traced onto the stone. The stone was cut and carved according to instructions.

9a   9b

9c

9d   9e

The mortar for both freestone and brick was "lime-sand mortar" and was popular among the colonies and Atlantic states until the late nineteenth century. It consisted of quicklime, sand, and water. Quicklime was created by heating crushed limestone to about 1650 degrees Fahrenheit. During this process, carbon dioxide was given off, creating quicklime (calcium oxide). When water was added to the quicklime, another process occurred called *slacking*. The lime would increase in volume, and heat would be given off.[7] Patrick Plunkett, master stone

9f   9g

It appears as though many small blocks of stone were placed around window openings at the ground floor of the White House. Tim Buehner illustrates the various steps of cutting and carving one solid piece of stone to give the illusion of a multistoned lintel section.
National Park Service

Birthstone of the White House and Capitol

carver and mason employed at the White House in the 1990s, said that the masons of old had to be careful with the heat produced by slacking. If slacked lime splattered on the skin or in the eye, it would seriously burn the worker.[8]

Slacking was usually done in pits. We know that this was the process used for the public buildings as Mrs. William Thornton, wife of the Capitol design winner, wrote in her diary in 1800 that old pits on the grounds behind the President's House contained water and rubbish.[9]

Decorative stone pieces were not attached to bare stone walls. For example, around windows of the White House are lintel stones. They appear as small pieces of stone inserted around windows, creating a decoration. However, portions of the face of the rock were cut away to create protruding lintel pieces. Then the entire block of stone would be placed around a window. One side of the stone fit snugly against the wall and the decorative section nestled against the window opening, creating the illusion of many stones rather than one solid piece. Skillful craftsmen accomplished this detailed work by careful planning and exact measuring.

Gorgeous roses, garlands, and ribbons that decorate the White House were not carved on the ground and attached to the walls. Instead, a large stone would be placed in the wall. A stone carver would mount the scaffolding and remove the unwanted stone. This created a flat wall with protruding decorations. Some common themes at the White House were chains, overlapping scales, acanthus leaves, mythical creatures, scrolls, medallions, cabbage roses, oak and acanthus leaves, roses, acorns, ribbons, bows, and swags.[10]

As previously mentioned, master craftsmen who created intricate decorative stonework and those skilled in cutting stone for accurate placement were difficult to find. Most masons came from Scotland. One group of six was hired from a Masonic lodge in Edinburgh. We know some cutters' and carvers' names by their signatures affixed to documents that were sent to the Commissioners or pay records. (* See note at end of chapter.) It appears a good number of these men worked at both construction sites during the initial 1790s construction period.

Exterior stone carving at the White House is considered by some architects to be the finest example of American eighteenth century stone carving. Surprisingly, there is no stone carving in the White House's interior. This, however, is not the case with the Capitol.

President Thomas Jefferson and Benjamin Henry Latrobe wanted stone carving to be superb in the interior of the Capitol. They first considered William Rush from America, but he dealt only in wood. Since no such craftsman could be found in the states, Latrobe wrote to Italy requesting sculptors. In his letter to a friend of Jefferson, dated March 6, 1805, he described Aquia stone and listed the qualities desired in a sculptor:

> The material in which this is to be cut is a yellowish sandstone of fine grain, finer than the peperino or gray sandstone used in Rome - the only Italian sandstone of which I have any distinct recollection. This stone yields in any direction to the chisel, not being in the least laminated nor hard enough to fly off (spall) before a sharp tool. It may, therefore, be cut with great precision. The wages given by the day to our best carvers are from $3 to $2.50, or about $750 to $900 per annum…. There are, however, other qualities which seem so essential as to be at least as necessary as talents. I mean good temper and good morals.[11]

In 1806, two young men, Giuseppe Franzoni and Giovanni Andrei, arrived in America. Andrei was Franzoni's brother-in-law and also his instructor. Their passage, as well as their families', was paid along with a salary of $85 per month. They initially shared a house provided for them near the Capitol. They were to work only for two years and then were to be sent back to Italy at the expense of the United States Government. Five years later Latrobe wrote that "…they have consented to remain an unlimited period on the same terms; and so unexceptionable has been their conduct, and their diligence, and so much is their talent required in the completion of the work, that no proposition has been made to or by them to put an end to their engagement."[12]

Latrobe first put Franzoni to work making an enormous eagle that he had envisioned for a frieze in the Hall of Representatives. Latrobe, however, had difficulty getting Franzoni to design an eagle appropriate for the Capitol of the United States. He wrote to his friend Charles Wilson Peale describing his problem.

> We have here two most capital Italian sculptors lately arrived. One of them is now modeling an eagle, but it is an Italian, or a Roman, or a Greek eagle, and I want an American Bald-eagle. May I therefore beg the favor of you to request one of your very obliging and skilful sons, to send me a drawing of the head and claws of the bald-eagle of his general proportions with the wings extended, and especially of the arrangement of the feathers below the wing when extended. The eagle will be fourteen feet from tip to tip of the wings, so that any glaring impropriety of character will be immediately detected by our Western members.[13]

Apparently Latrobe was pleased by Franzoni's finished product, for on December 19, 1806, he wrote to the Italian who was instrumental in obtaining Andrei and Franzoni:

> Our Italian sculptors continue to give us the utmost satisfaction…Franzoni has been only engaged with eagles. He has finished a colossal Eagle in the Frieze of the Entablature of the Hall of Representatives 12'6" from wing to wing, and is now

Birthstone of the White House and Capitol

The cornstalk, or corncob, capitals were designed by Benjamin Latrobe and modeled by Giuseppe Franzoni. Latrobe wrote Jefferson, "This capital, during the summer session obtained me more applause from the members of Congress than all the works of magnitude...." They are located on the first-floor vestibule of the Old Supreme Court Chamber.
Architect of the Capitol

> engaged on a free eagle, also colossal for the Gate of the Navy Yard. This promises to be the most spirited Eagle I have seen in sculpture either modern or antique.[14]

In the House of Representatives, Franzoni carved other rich decorations that unfortunately did not survive the 1814 fire.

Giovanni Andrei modeled and carved capitals (tops) for the columns in the Senate stair vestibule that Latrobe had designed. They were uniquely American, for instead of the traditional acanthus leaves he used corn. Andrei was a slow but conscientious workman. Latrobe wrote to a commissioner that "[Andrei] is the slowest hand ever I saw, especially in modelling, and in fact our clay models of his work have cost more than the same thing in marble."

The corn capitals won acclaim for Latrobe and Andrei. A woman who visited the Capitol wrote in *Domestic Manners of the Americans*, "In a hall leading to some of these rooms the ceiling is supported by pillars, the capitals of which struck me as peculiarly beautiful. They are composed of the ears and leaves of Indian corn, beautifully arranged and forming as graceful an outline as the acanthus itself." In 1809 Latrobe shipped a model of the capital to Thomas Jefferson, who had recently retired to Monticello. (Jefferson used it to support a garden sundial.) Latrobe wrote the following in a letter accompanying his gift:

> This capital, during the summer session obtained me more applause from the members of Congress than all the works of magnitude.... They called it the Corn Cob Capital, whether for the sake of the alliteration I cannot tell, but certainly not very appropriately (for the capital uses the full ears, not the cob).[16]

Andrei and Franzoni proved invaluable to Latrobe. In 1812, six years after their arrival in America, Latrobe wrote to Jefferson about these fine Italian stone carvers:

> Andrei, the Sculptor of decoration, is one of the most estimable characters I know, and in his manners, language, and feelings a perfect Gentleman. Franzoni, who employs chisel only in figures, is not the same sort of being. He is full of genius, but his habits and manners are those of a Mechanic.[17]

The rotunda walls, constructed of Aquia freestone, contain bas-reliefs.
Architect of the Capitol

After the Capitol's destruction in 1814, carving was done for restoration. Additional carving of Aquia stone in the interior of the Capitol Building was accomplished in 1823. Over each of the four principal doorways in the rotunda is a large, vertical panel with figures in low relief. The principal carvers were Antonio Capellano and Enrico Causici of Italy and Nicholas Gevelot of France. They received $3,500 for each relief.[18] The sculptors were classically trained and worked primarily in marble and were not used to carving American figures. Their bas-reliefs have been targeted for ridicule throughout the ages. For example, some critics commented that Pocahontas' face and headdress were Grecian.

The panel, the *Rescue of John Smith by Pocahontas*, was carved by Antonio Capellano. Rembrandt Peale wrote an article for *The Crayon* in 1856 that contained an amusing anecdote about Capellano:

Birthstone of the White House and Capitol

Rescue of John Smith by Pocahontas. Architect of the Capitol

He was a most industrious man and so devoted to his marble that he could not spare an hour to learn either French or English; and his wife who had joined him from New York, told me that she believed that he would turn to stone himself. Fifteen years after this (in 1830) I was surprised one fine afternoon in the Boboli gardens at Florence on being accosted by a well dressed Signor with his gay wife and five fine children. It was Capellano, who…informed me that having made money enough in America, he had bought "uno picollo palazzo" to enjoy the remainder of his days in his native city.[19]

Enrico Causici created two panels, *The Landing of the Pilgrims* and the *Conflict Between Daniel Boone and the Indians*. The story is told that in the early 1900s, twenty Winnebago Indians, dressed in full war regalia, visited Washington and, while touring the Capitol viewed the panel depicting Boone fighting an Indian brave. After examining the bas-relief, the group of Native Americans rapidly exited the scene with loud cries and war whoops. (Causici is not primarily known for his work in the Capitol, but rather for impressive sculptures in Baltimore.)[20]

The fourth panel, *William Penn's Treaty with the Indians* was carved by Nicholas Gevelot. Not much is known about this Frenchman, but on November 28, 1826, John Quincy Adams wrote about his visit to the Capitol and meeting Gevelot.

> A French workman in sculpture, engaged upon a bas-relief of Penn's Treaty, came and asked me to go up on his scaffolding and view his work; which I did. But all the bas-reliefs in the rotunda are execrably bad. I went up likewise within the scaffolding to the pediment, where Persico was at work. One of his three figures is nearly finished, and I think the design when completed will be good.[21]

Horizontal bas-reliefs, carved in 1827, are also located on the rotunda walls. Each consists of a floral wreath with relief portrait. The heads are of Columbus, Sir Walter Raleigh, La Salle, and Cabot.[22] The two Italians, Capellano and Causici, were given $9,000 for completion of the freestone reliefs.[23]

Cutting and Carving Aquia Stone

*Conflict between Daniel Boone and the Indians.
Architect of the Capitol*

William Penn's Treaty with the Indians.

The exterior of the Capitol, unlike the White House, did not have intricate stone carving. Carving, however, could be found on the north and south wings in the decorative capitals of the pilasters. Latrobe gave an explanation for the lack of intricate carving in a letter he wrote in February of 1811:

Birthstone of the White House and Capitol

> It has been my wish to diminish the quantity of external sculpture as much as possible as may be seen or compared [in] the two wings both on account of its cost, and because our freestone is not as durable when nicely carved as might be wished.[24]

Some carving on the east pediment over the exterior entrance to the Rotunda was created by Italian Luigi Persico. Supposedly, drawn to America in order to obtain wealth, he ended up in Philadelphia painting miniatures and teaching drawing. A chance contest win creating a medal led him to be chosen for work at the Capitol. In a letter dated June 22, 1825, Architect of the Capitol Charles Bulfinch mentioned that work on the portico had slowed down due to delays in receiving the freestone. He also described the sculptured reliefs:

> After several attempts, the following has been agreed upon; a figure of America occupies the centre, her right arm resting on the shield, supported by an altar or pedestal bearing the inscription, June 4, 1776, her left hand pointing to the figure of Justice, who with unveiled face, is viewing the scales, and the right hand presenting an open scroll inscribed Constitution, March 4, 1789; on the left of the principal figure is the eagle, and the figure of Hope resting on her anchor, her face and right hand uplifted, - the whole intended to convey that while we cultivate Justice we may hope for success. The figures are bold, of nine feet in height, gracefully drawn by Mr. Persico, an Italian artist. It is intended that an appropriate inscription shall explain the meaning and moral to dull comprehensions.[25]

*Collen Williamson was master mason at both sites until fired in 1795. Henry Edwards, James Dixon, and Joseph Huddleston were foremen of stone carvers at the President's House at various times from 1794 to 1798. Robert Vincent, James Dougherty, David Cummings, Robert Tolemy, William Timmener, John Davidson, and John Hogg were

Horizontal bas-relief of John Cabot.
Architect of the Capitol

Cutting and Carving Aquia Stone

This pediment carving by Luigi Persico is located over the Central Portico on the East Front of the Capitol. Originally carved of sandstone, it has since been duplicated in marble.
Architect of the Capitol

also listed as stone carvers. Evidently sometime later they shifted over to the Capitol site. According to Robert Kapsch in his dissertation, *Labor History of the Construction and Reconstruction of the White House, 1793–1817*, "At no single time were there more than five stone carvers working on the President's House."[26] Some stone cutters were Joseph Huddleston, Hugh Sommerville, James Somervelle, William Symington, Alexander Wilson, William Bond, Francis Reid, James McIntosh, Alexander Reid, James Reid, John Williamson, James Dougherty, and Andrew Shields. Several of the cutters also worked at both sites.[27]

15

War of 1812

The War of 1812 technically began when President James Madison signed a declaration of war on June 18, 1812. However, there were many events, dating as far back as 1802, that led to the outbreak of a second war against the British Empire. Once declared, fighting began in the U.S. near the Canadian border and inside Canada. At one time in the hostilities, the American troops burned York, in Upper Canada. By 1814 the complexion of the war changed in that Napoleon abdicated. This freed British forces previously at war with the French to deploy to America. Fighting occurred near Niagara and Lake Champlain. British forces from Bermuda came to Chesapeake Bay and defeated Americans at Bladensburg. They then marched on Washington, D.C., in retaliation for the Americans attacking York.[1]

Dolley Madison's letters leave a picture of impending events. She writes to her cousin, Edward Coles, on May 12, 1813, a year prior to the initial attack.

> ...And now if I could I would describe to you the fears and alarms that circulate around me.... One of our generals has discovered a plan of the British,-it is to land as many chosen rogues as they can about fourteen miles below Alexandria, in the night, so that they may be on hand to burn the President's house and offices....[2]

Birthstone of the White House and Capitol

Fifteen months later, on August 23, 1814, Mrs. Madison wrote to her sister, Anna:

> Dear Sister, - My husband left me yesterday morning to join General Winder. He inquired anxiously whether I had courage or firmness to remain in the President's house until his return on the morrow, or succeeding day, and on my assurance that I had no fear but for him, and the success of our army, he left, beseeching me to take care of myself, and of the Cabinet papers, public and private…I have pressed as many Cabinet papers into trunks as to fill one carriage; our private property must be sacrificed, as it is impossible to procure wagons for its transportation….[3]

Dolley Madison continues her letter to her sister on the next day, August 24:

> Three o'clock. - Will you believe it, my sister? we have had a battle, or skirmish, near Bladensburg, and here I am still, within sound of the cannon! Mr. Madison comes not. May God protect us!…Our kind friend, Mr. Carroll, has come to hasten my departure, and in a very bad humor with me, because I insist on waiting until the large picture of General Washington is secured, and it requires to be unscrewed from the wall. This process was found too tedious for these perilous moments; I have ordered the frame to be broken, and the canvas taken out. It is done!…And now, dear sister, I must leave this house….[4] (See page 24 for painting she saved.)

After Mrs. Madison left, the President returned for a short time. He did not eat the dinner that was prepared but had only a bit of wine. Following his departure, Madison's French porter ironically locked the house.[5]

British troops arrived in the district about 7:30 at night. They marched to the Capitol, where they followed their orders to "destroy and lay waste." It has been said that British Rear Admiral Cockburn led troops into the House Chamber and mounted the Speaker's chair and asked the question, "Shall this harbor of Yankee democracy be burned?" to which the troops eagerly replied, "Aye!" Books from the Library of Congress, tar barrels, and pieces of furniture were stacked in piles creating fuel for a blaze.[6] The French minister, Louis Serurier, who was staying in Washington described the burning of the Capitol shortly after 9 p.m. "I have never beheld a spectacle more terrible and at the same time more magnificent."[7]

This picture from an English history book shows the British burning D.C. in 1814.
Rapin's History of England, Library of Congress

Following the torching of the Capitol, Cockburn marched down Pennsylvania Avenue to the Presidential Mansion with more than 150 seamen.[8] Breaking open the door, they found a vacant White House. One of the officers remembered, "[W]e found a supper all ready which many of us consumed…and drank some very good wine also."[9] While some dined, others ransacked the mansion and prepared piles of furniture, books and provisions in the rooms. Mrs. William Thornton, wife of the designer of the Capitol, recalled to a friend that "50 men, sailors and marines, were marched by an officer, silently thro' the avenue, each carrying a long pole to which was fixed a ball about the circumference of a large plate." After those British inside exited the building, the torches, made of rags soaked in oil, were hurled through the broken out windows. Then "an instantaneous conflagration took place and the whole building was wrapt in flames and smoke. The spectators stood in awful silence, the city was light and the heavens redden'd with the blaze…."[10]

The Aquia stone walls of the Capitol and President's House were heated, only to be cooled by a violent rainstorm.
The Kiplinger Washington Collection

The National Intelligencer later reported that Cockburn was "…exhibiting in the s[t]reets a gross levity of manner, displaying sundry articles of triffling value of which he had robbed the president's house."[11]

U.S. Attorney General Rush wrote a first-person account of the fire:

> I have indeed, to this hour, the vivid impression upon my eye of columns of flame and smoke ascending throughout the night…from the Capitol, President's house, and other public edifices, as the whole were on fire, some burning slowly, others with bursts of flame and sparks mounting high up in the dark horizon….[12]

Both the White House and the Capitol burned, turning the night sky red. The Aquia stone of both structures got extremely hot, only to cool down with a sudden thunderstorm. The next day more bad weather attacked the already damaged shells. An Englishman described the event:

Birthstone of the White House and Capitol

…the most tremendous hurricane ever remembered by the inhabitants broke over Washington the day after the conflagration. Roofs of houses were torn off and carried up into the air like sheets of paper, while the rain which accompanied it was like the rushing of a mighty cataract rather than the dropping of a shower. This lasted for two hours without intermission, during which time many of the houses spared by us were blown down.[13]

RUINS AND REBUILDING

Two days after the burning, President Madison and his cabinet returned to Washington. A citizen, after viewing the ruins, wrote that the only things that remained were "…unroofed, marked walls, cracked, defaced, blackened with the smoke of fire."[14] There was even graffiti on the walls of the Capitol, done by a party that blamed Madison with its destruction. The walls read, "George Washington founded this city after a seven years' war with England - James Madison lost it after a two years' war."[15]

Margaret Bayard Smith, whose husband established the *National Intelligencer*, wrote her sister:

> The poor capitol! …We afterwards look'd at the other public buildings, but none were so thoroughly destroy'd as the House of Representatives [south side of the Capitol] and the President's House. Those beautiful pillars in that Representatives Hall were crack'd and broken, the roof, that noble dome, painted and carved with such beauty and skill, lay in ashes in the cellars beneath the smouldering ruins, were yet smoking. In the P.H. [President's House] not an inch, but its crack'd and blacken'd walls remain'd….[16]

The destruction of the two most important buildings in the nation's capital was denounced not only in America but also in England. *The London Statesman* commendably wrote, "The Cossacks spared Paris, but we spared not the capitol of America."[17]

Since the Madisons could not return to the White House, they stayed for a month with his brother-in-law, Richard Cutts.[18] Then on September 9, 1814, the *National Intelligencer* wrote that "the President will occupy Colonel Tayloe's large house…."[19] The house to which the newspaper article referred was known as The Octagon. Located two blocks from the President's Mansion, the home was built on a triangular lot. Construction was started in 1799, prior to completion of either the White House or Capitol. John Tayloe III and his wife selected Dr. William Thornton, first Architect of the Capitol, to design this winter town home. Their friend, President Washington, had encouraged them to build in the nation's new capital city. Tayloe was a wealthy tobacco planter and breeder of race horses. His family's home was Mount Airy in Richmond County, Virginia, thought by many historical architects to be the most ambitious house built during the Colonial period in Virginia.[20] Although constructed of brownish-gray sandstone, Mount

Airy had white Aquia stone trim. The Octagon, made with red bricks purchased within the city, also used Aquia stone for trim.

As the Madisons settled into The Octagon, congressmen returned to the city and were shocked to see the damage. Representative Jonathan Roberts wrote, "The ruins of the public edifices is more complete than I had anticipated."[21]

The homeless Congress met in the Patent Office Building. It was the only government building to escape destruction. The building was saved by Dr. William Thornton, who convinced the British that burning it would be a "barbarous" act since it contained only private property.[22]

The first thing Congress had to face in their cramped quarters was whether to relocate the nation's capital to another city. Northerners saw this as the perfect opportunity to move out of the District of Columbia. City officials in Philadelphia promised suitable accommodations. John Quincy Adams wrote his wife, "The removal of the seat of government necessitated by the event may prove a great benefit rather than a disadvantage to the nation."[23] Southerners and many Washingtonians, however, did not wish to abandon the city. Margaret Bayard Smith, a leader of Washington society, wrote:

> Oh that I a feeble woman could do something! This is not the first capital of a great empire, that has been invaded and conflagrated; Rome was reduced still lower by the Goths of old, than we are, and when its senate proposed removing the seat of government, they were answered, Romans would never be driven from their homes, Rome should never be destroy'd. May a Roman spirit animate our people, and the Roman example be followed by the Americans.[24]

The United States' victory at the Battle of New Orleans by General Andrew Jackson changed the attitude of Congress. With renewed confidence, it decided to rebuild the city. They also authorized the purchase of Thomas Jefferson's 10,000-volume library since the British had burned the volumes housed in the Capitol.[25]

The victory in New Orleans, however, had no effect on the war, as peace had already been established two weeks earlier in Ghent, Belgium, on December 24, 1814. The peace treaty had to be approved by the United States Senate and President Madison. On February 17, 1815, the document was delivered to the President in The Octagon and signed there by him. The Madisons lived in The Octagon for seven months and then moved into a smaller house at Pennsylvania Avenue and Nineteenth Street. They never did return to the White House, as it was rebuilt during the remaining months of his presidency.[26]

16

Rebuilding

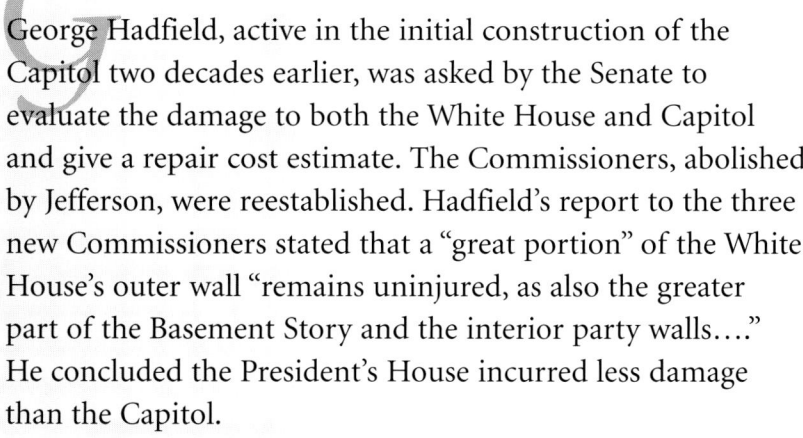

George Hadfield, active in the initial construction of the Capitol two decades earlier, was asked by the Senate to evaluate the damage to both the White House and Capitol and give a repair cost estimate. The Commissioners, abolished by Jefferson, were reestablished. Hadfield's report to the three new Commissioners stated that a "great portion" of the White House's outer wall "remains uninjured, as also the greater part of the Basement Story and the interior party walls...." He concluded the President's House incurred less damage than the Capitol.

In his report Hadfield estimated that the cost of restoring the public buildings to their former state would require $692,000:

> For taking down all the ruinous parts of the Capitol, and removing all the ruins and rubbish; scaffolding and labour included $20,000
> Ditto for the President's house $25,000
> Ditto for the two Offices $9,000
>
> For restoring the Capitol to its former state, workmanship, labour and materials included $340,000
> Ditto for the President's house $270,000
> Ditto for the two Offices $27,000
> Total of the expence $692,000

George Hadfield[1]

Birthstone of the White House and Capitol

Margaret Bayard Smith wrote of the President's House, "In the P.H. not an inch, but its crack'd and blacken'd walls remain'd...."
William Strickland, Library of Congress

Congress, however, appropriated only $500,000 for restoration, $192,000 less than Hadfield had estimated. President Madison asked the Commission to rebuild and "not deviate from the models destroyed."[2]

Congressmen did not like meeting in the Patent Office Building. Daniel Webster said that their new quarters were "confined, inconvenient, and unwholesome."[3] They then moved into a hurriedly constructed building and rented it from private citizens for four years. Known as the Brick Capitol, it was located on the site of the present-day United States Supreme Court Building.[4]

No building restoration was accomplished during the winter of 1814–1815 because of severe weather. But during the spring, Benjamin Henry Latrobe, second Architect of the Capitol, returned to rebuild the Capitol and examine the President's House. He was reappointed by President Madison on April 6, 1815. During his departure from Washington, Latrobe had been in Mississippi designing steamboats, in New Orleans

planning waterworks, and in Pittsburgh where he broke off his connection with Robert Fulton's failing steamboat enterprise. After returning to Washington and viewing the destruction, Latrobe wrote to his wife:

> …the ruins of the Capitol, which, I assure you, is a melancholy spectacle. There is no describing it. However, many important parts are wholly uninjured, and what particularly is gratifying to me, the picturesque entrance of the house of Representatives with its handsome columns, the Corn Capitals of the Senate Vestibule, the Great staircase, and all the Vaults of the Senate chamber, are entirely free from any injury which cannot be easily repaired. Some of the Committee rooms of the southwing are not even soiled, but in general the wood work is burnt in patches.[5]

Later, Latrobe wrote to Thomas Jefferson and gave him details about the fateful night when the British burned the Capitol. He said that they had attempted to kindle fires in most rooms. "A man with an axe chopped the wood work, another followed behind and brushed on…" rocket powder and a third lit it. "In the Clerks office the desks and furniture and the records supplied a more considerable mass of combustible materials than there was elsewhere: and the fire burnt so fiercely that they were obliged to retreat and leave all the rooms on the West side entirely untouched, and they are now as clean and perfect as ever."

Margaret Bayard Smith wrote of the Capitol ruins, "Those beautiful pillars in that Representatives Hall were crack'd and broken, the roof, that noble dome, painted and carved with such beauty and skill, lay in ashes in the cellars beneath the smouldering ruins."
George Munger, 1814 or 1815. The Kiplinger Washington Collection

Latrobe explained that the British had found that the roof of the House of Representatives was indestructible. So they assembled all the stages and seats of the galleries, which "were of timber and yellow pine. The Mahagony furniture, desks, tables and Chairs were in their places. At first they fired Rockets through the Roof. But they did not set fire to it: they sent men on to it, but it was covered with Sheet Iron. At last they made a great pile in the Center of the room of the furniture, and retiring, set fire to a large quantity of Rocket stuff in the middle." He said that the fire was so intense the glass of the lights was melted, as he found lumps of molten glass "weighing many pounds." Latrobe continued and discussed the fire's effect on the Aquia stone. "The stone, is like most freestone, unable to resist the force of flame. But I believe no known material could have withstood the effects of so sudden and intense a heat."[6]

Thirteen days after he was hired, Latrobe wrote to the Commissioners about his plans for the south wing of the Capitol:

> The expedition with which the work will proceed depends mainly upon the supply of materials, and the facility with which workmen may be procured.... The inter-ior of the Hall of Representatives will be carried up entirely in freestone if the room should be rebuilt exactly upon the former design. The stone is still in the Quarries and the Quarries are not yet opened.... No time therefore is to be lost in making arrangements with the proprietors of the Quarries on Acquia.[7]

Latrobe continued by writing that he wanted an "...enclosure of ground sufficient for the Stone cutters Yard in front of the South wing...."[8] and ended by expressing concern about the stone and workmen:

> ...after consultation with Mr. Blagden...it is my opinion that there will be a great difficulty in procuring a sufficient number of Stonecutters to proceed with all the contemplated buildings at once, but none in the supply of Stone and other Materials, and that for freestone the principal dependence will be to be placed on the quarries at Acquia."[9]

A week later, Latrobe wrote the Commissioners, again stating that using Italian marble for the column capitals in the Hall of Representatives would be better. He could obtain capitals from Italy fully carved and ready for installation only "more than $\frac{2}{3}$ds as much as they have cost us in the brittle freestone of Acquia."[10]

After visiting Loudoun County, Virginia, and viewing marble outcroppings, Latrobe believed that this stone could be used for column shafts in the "Hall of Representatives."[11] He later looked at the red sandstone at Seneca Falls, Maryland. Although red, it was harder than Aquia stone and would "undoubtedly furnish all our steps, pavement, and Ballusters as well as cornice Mouldings...."[12]

Meanwhile, at the President's House, James Hoban was asked to return. At age fifty-seven, he was asked to rebuild the structure to its original glory. Since his departure from the White House he had remained in Washington with his wife and nine children and had invested his savings in real estate.[13] After the walls were cleaned and the interior swept out, it was discovered that Hadfield's appraisal of the damage was too conservative. The Aquia stone walls were standing, but sections had been damaged and split due to the intense heat of the fire and the sudden cooling from the thunderstorm.

Hoban set to work, reopening the government's quarry on Aquia and making arrangements for reconstructing the President's House. His letter to the Commissioners stated:

> The Stone cutters Shed is now ready for operations in that line and a Work shop for the carpenters is progressing, which will be put up between the Presidents house and the Treasury Office, and will ansr. for the work of the fire proof also....[14]

Besides "old-timers" Hoban, Latrobe, and Hadfield, the Commissioners recalled George Blagden. His new job would be inspecting stone and as superintendent of stone cutters and setters at the Capitol and the President's House. Robert Brown, who had worked at the President's House years before, was rehired as foreman of the stone cutters and carvers. Italians Carlo Franzoni and Giuseppe Valaperta were borrowed from the Capitol to replace exterior carving.

Robert Kapsch, former chief of the Historic American Buildings Survey, wrote in his dissertation, *The Labor History of the Construction and Reconstruction of the White House, 1793–1817*, that one of the reasons "that the building process of the reconstruction was very similar to the initial construction was that the damage done to the buildings was so severe. Although some parts of the buildings could be used, most of the restoration involved totally new construction. This new construction was planned and executed along the same lines, using the same building materials, as the original construction."[15]

This time, construction was easier as roads and transportation had improved during the intervening twenty years. The Commissioners asked for bids for stone, lime, bricks, copper, marble, iron, yellow pine plank, and sand. For example, in February 10, 1816, they wrote in the *Daily National Intelligencer* that they needed "2,400 tons of freestone, of the first quality to be approved of by the agent of the Commissioners on delivery, and of the dimensions which may be specified."[16]

As with the initial construction, skilled laborers were difficult to find. Advertisements, such as the following, were sent from Richmond to Boston:

Commissioner's
Buildings, April 1st, 1816
STONE CUTTERS WANTED

> The Commissioners of the Public Buildings hereby give notice, that they will give the following wages to any number of Stone Cutters, from this day until the first of October next, rating the day at ten hours: To every Stone Cutter, a first rate workman, $250 cents per day.
>
> To all others in the same proportion, according to the quality and quantity of work they can perform per day. A large number of Stone Cutters are now wanted, and will be employed at these prices, until the first of October, after which, liberal wages will be given during the winter, and constant employment until the Public Buildings are finished.[17]

This time there was not the intensive program of recruiting overseas workers. However, some workers were met at the docks in New York and were recruited on the spot. The Commissioners guaranteed positions and pay for passage. The passage payment came out of the workers' wages. But, if skilled workers came from cities within the United States, like Baltimore, the Commissioners would "pay their passage in the cheapest mode to this place."[18]

During initial construction of both federal buildings, there was great debate over workers' pay, by piece or by the day. During the rebuilding most workers were paid by the day.

In October 1815, stone cutters at the Capitol requested an increase in wages. The Commissioners delivered a very stern reply stating that they were already given high wages compared to others doing similar work.[19]

When a comparison is made of the wages given during the initial construction with that of rebuilding, it is evident that there was a substantial increase. For example, during the construction period, the stone cutter foreman received $2 per day. During reconstruction this was increased to $3.75 daily. Likewise, stone cutters' pay increased from $1.25–$1.33 per day to $2.50–$2.75 per day.[20]

Pay records indicate that slaves and apprentices, commonly used during construction in 1791–1800, were not used during the rebuilding program. Most likely, common laborers were easier to obtain.

Just as the Commissioners were abolished during the Jefferson administration, those for reconstruction were abolished by James Monroe on April 29, 1816, and replaced by Colonel Samuel Lane, an old friend of James Monroe. Lane was a small man and had been partially crippled by a musketball wound. He liked to say it was a war injury, but in actuality it was due to an accident.[21] He rode around the district inspecting reconstruction in a two-wheeled horse cart and made people come to him.[22]

Prior to tearing down damaged sections of Aquia stone walls, stone masons worked for a year in their sheds preparing stone for reconstruction. Reconstruction of the White House basement was unnecessary. Collen Williamson's work and the solid Aquia stone walls had survived the ravages of war.

Latrobe started restoring the Capitol. He used Aquia stone for most of the repairs and introduced new building materials that he had found, such as Italian marble, Loudoun County marble, and Seneca stone. He found no fine Aquia stone "of a texture for the finer works of the buildings" so he recommended marble.[23] However, the Aquia quarries were actively trying to fill orders for general reconstruction. Latrobe's son, Henry S. B. Latrobe, wrote to his father in August of 1817, asking if he would recommend getting sandstone from Cuba or Aquia for a lighthouse he was building in New Orleans. Latrobe replied, "As to the idea of getting stone from Acquia I hardly know what to say to it. Many quarries are opened, but all are occupied, and Stone is from 10 to 12.$50 per Cubic foot."[24]

Tobacco capital.
Architect of the Capitol

As Latrobe had mentioned to his wife, his Aquia stone corncob capital columns in the Capitol survived the fire. During the rebuilding period he designed other capitals to crown the columns of the oval lobby on the upper floor. These capitals were also carved from freestone and had a tobacco leaf motif. Today they are found in the area called Hall of Columns or "Tobacco Hall." These were modeled and carved by another Italian, Francisco Iardella. Just as he had done before for the corn capital, Latrobe sent a model to Jefferson. In an accompanying letter he stated that these capitals were not as effective as the corncob capitals and suggested that the President stain the leaves pale umber to bring out the flowers, for this is what was going to be done at the Capitol.[25]

Birthstone of the White House and Capitol

The tobacco capitals were designed by Latrobe after the British burning in 1814. They were carved out of freestone by Francisco Iardella. Today columns with tobacco capitals can be found in the lobby of the small rotunda on the second floor and in the Hall of Columns located on the first floor of the House wing.
Glenn Brown, Architect of the Capitol

In 1816, the Commissioners set a standard workday of ten hours for all the workers except for the laborers, who would work the whole day, or from "sunrise to sunset" as before.[26] Naturally, this did not set well with the laborers, and they protested. Latrobe wrote to Samuel Lane, "There is at present a total desertion of all the Work by our Laborers, in order to force a regulation respecting their time…."[27]

On October 10, 1816, James Hoban wrote the Commissioners that "Having commencd. the raising of the Roof of the President House, and having got one of the Principal Rafters in its Place on the Building…" that he wanted to have a celebration for about one hundred men. He requested money for refreshments and reminded them that such a party occurred after the original roof was raised. Peter Lennox, foreman of the carpenters, ordered the refreshments consisting of 31 gallons of whiskey, 161 loaves of bread, 42 pounds of cheese, 8 bags of crackers and 11 pounds of sugar![28]

By April of 1817, President Monroe ordered that the buildings be completed with "dispatch."[29] Hoban hired more men.

Benjamin Henry Latrobe worked for two years restoring the exterior of the Capitol as well as its interior. He carefully removed all the damaged stone and helped rebuild the House of Representatives, the chamber he had originally constructed in 1807. Today, known as Statuary Hall, one can still see the warm tones of the freestone walls along with the rich, polished marble columns. The [Old] Senate Chamber, which was also destroyed by the fire, was enlarged by Latrobe to its present dimensions. Latrobe also completed some drawings for rebuilding the President's House. Despite his efforts at restoration, Latrobe resigned on November 20, 1817, in dispute over his authority at the Capitol.

> *Samuel Lane made life miserable for Latrobe by continually insinuating to President Monroe that Latrobe was mismanaging the rebuilding at the Capitol.*

Samuel Lane made life miserable for Latrobe by continually insinuating to President Monroe that Latrobe was mismanaging the rebuilding at the Capitol. Mrs. Latrobe related an incident that was the deciding factor for her husband's resignation. One day Samuel Lane publicly reprimanded the architect before the President. Mrs. Latrobe said that her husband "seized him by the collar, and exclaimed, 'Were you not a cripple I would shake you to atoms, you poor contemptible wretch. Am I to be dictated to by you?' The President said, looking at my husband, 'Do you know who I am, Sir?' 'Yes, I do, and ask your pardon, but when I consider my birth, my family, my education, my talents, I am excusable for any outrage after the provocation I have received from that contemptible character.'"[30]

Birthstone of the White House and Capitol

This rounded dome graced the Capitol from 1825 to 1856. Designed by Charles Bulfinch, it was made of wood and covered with copper. Up until the middle 1800s, buildings were recorded by artists' renderings, but in 1846 John Plumbe created this daguerreotype, the first photographic-type image of the Capitol.
Library of Congress

After Latrobe's resignation, Charles Bulfinch was appointed Architect of the Capitol. Born in Boston, this fifty-four-year-old gentleman had the distinction of being the first American-born architect in charge of the Capitol building. He was Harvard-educated, studied in Europe, and spent many years practicing his craft in America.[31]

On August 24, 1818, another cornerstone was laid, this time for the Capitol Rotunda. Coincidentally, the date was four years to the day of the burning of the Capitol by the

British.[32] Bulfinch started work on the Rotunda as well as making sure that the Senate and House chambers were structurally correct. Work progressed sufficiently by 1819 that Congress was able to return.

The Rotunda, or "rotunda" as it was called in much of the nineteenth century, was now ready for a dome. Thornton envisioned a low and graceful dome, like that of the Pantheon in Rome. Bulfinch wished to carry out Thornton's design, but President Monroe and his Cabinet said that it should be higher. Bulfinch carried out the President's orders and made a higher dome of wood covered with copper. William Thornton thought it looked "ridiculous." He wrote to a friend that it reminded him of a large sugar dish between two tea canisters.[33] Anne Newport Royall, America's first woman journalist, described the dome on her visit to the Capitol in the mid-1820s. "The great centre dome in shape resembles an inverted wash-bowl; only magnify a wash-bowl to the size of ninety-six feet in diameter, and you will have the correct idea of its figure."[34]

17

Capitol's Completion and Quarries' Closure

Columns were added to the east front of the Capitol before its completion. Dr. William Thornton was the first to envision columns. His winning entry, reminiscent of the Pantheon of ancient Rome, featured a classical entrance with a grand portico. The triangular portico, or porch, was to be held up with stately Corinthian columns. Some trained architects criticized his ambitious plan for this noble structure, but Thornton remained true to his belief that the Capitol of the new nation should echo the architecture of ancient Rome, thus giving dignity to the building. Once he wrote, "I worked day and night at [designing] the Capitol. I finished, and obtained the prize against a world of competitors: some regularly bred architects. I went at once to the highest order - viz. the Corinthian. I was attacked by Italian, French and English - I came off, however, victorious."[1]

A Corinthian column is distinguished from other types by its ornate decoration. The ancient Greeks developed three distinctive types of columns—the Doric, Ionic, and Corinthian orders. Each shared some principal sections. A *shaft*, or vertical piece, usually rested upon the *base,* which in turn rested upon the *pedestal.* On the top of the shaft was an enlarged section known as a *capital.* The Corinthian column was graceful. Its shaft was slender, and the capital was shaped like an inverted bell and decorated with volutes (spiral scroll-like ornaments) and acanthus leaves.[2]

Thornton's original design contained twelve columns. Benjamin Latrobe revised Thornton's design. He added depth and width to the portico and increased the dozen columns to twenty-four. After Latrobe's resignation, Charles Bulfinch reviewed his predecessor's plans and decided on the twenty-four columns Latrobe suggested.[3]

Since these columns were going to be at the impressive East Front entrance, it was thought that the shaft portion should be made out of one solid piece of stone. A monolithic shaft would express more perfection and power than one constructed from several pieces of stone like those shafts on the West Front. In order to obtain such huge pieces of sandstone, Joseph Elgar, the Commissioner of Public Buildings, needed to engage contractors. Thomas Towson, who now ran the island quarries, wrote to Elgar on October 7, 1823: "Agreeably to your instructions I have made an experiment on the Island Quarries, the result of which leaves no doubt in My Mind, but the collumns for the Portico of the Capitol May all be quarried there." Towson explained that it would be advisable to quarry them there "on account of its convenience to the water."[4]

Elgar replied to Towson three weeks later that he and George Blagden, superintendent of stone work and quarries, would visit the island on November 22. Towson was pleased and responded by giving a description of the rock that he thought they might use. He believed the stone was "of excellent quality" and gave dimensions of blocks that he thought he could quarry. They were "from 35 to 40 ft." long and "about 9½ ft. thick."[5] Even before Blagden and Elgar arrived, Towson wrote again and said that "I expect to have one column shaft blocked out the present week." He did not think he would have any difficulty quarrying the stone but wrote, "I apprehend more difficulty in preparing a road [to take them to water] in time, to deliver the columns, before winter setts in…."[6]

In the meantime, Elgar had to arrange transportation for the stones that Towson would obtain from the quarry. He contacted Withers Waller of Stafford who wanted $666 for the transportation of the twenty-four shafts. Elgar was appalled at his request and said that the large amount was "a Sum more than sufficient to build a scow for their purpose." He informed Waller that he would pay only $2 per ton.[7] A month later, Waller and his partner, Morton, decided to take Elgar's offer.[8]

Elgar wrote to Towson in February of 1824 and informed him of Waller and Morton's intentions to transport the shafts. He also said, "The Committee of Congress have reported in favor of the Portico to the President's House. If the Appropriations should get through, we shall want 8 Columns and other stone for that."[9]

Capitol's Completion and Quarries' Closure

This photograph, taken in 1860, shows cutters working on marble columns. Since photographs were not available during the original construction of the Capitol, one can only imagine that the workplaces looked similar in the early 1800s when the Aquia stone was cut.
Architect of the Capitol

The next month Thomas Towson wrote Elgar and appeared delighted to inform him that "I have now the Sixth Column quarryed and the 7, 8, & 9 in considerable forwardness, and I am happy to inform you that my Quarrys still present a very favorable aspect."[10]

In 1824, sixty to seventy stone cutters decided to have a parade to celebrate the Fourth of July, so they placed an advertisement in the *Washington Gazette*. A procession would display the "...cutting of a Corinthian capital, intended to crown one of the Eastern front columns, and a key stone, for one of the arches." They mentioned those facts "...to shew the public spirit likely to manifest itself on our approaching National Anniversary."[11]

The celebration was a huge success. The July 6 *Washington Gazette* reported that "...President's [Monroe's] carriage was followed by those of Mr. Secretary Adams and Mr. Secretary Calhoun, and the Foreign Ministers." The Typographical Society had a printing press on a platform that printed copies of the Declaration of Independence as their float rolled down the avenue. Copies were then distributed to the spectators. "The Stone Cutters, with a large Corinthian Capital, wrought in freestone, on a moveable stage, gave a fine specimen of their art."[12]

After the parade, problems arose with the stone that had already arrived in the district. Unfortunately for Thomas Towson, who worked so hard getting the shafts out of the quarry, George Blagden examined the stone blocks and claimed that he saw

Birthstone of the White House and Capitol

Fluted cast iron columns were hauled by six horses during the construction of the dome in 1857. During the construction of the East Front in the 1820s, an Aquia stone column shaft was placed on a wagon and pulled by one hundred men.
Architect of the Capitol

cracks and flaws. Blagden notified Elgar and said he felt that a third person should be involved, one who was familiar with the island and its stone. This person could determine if there was enough good stone to get the monolithic columns out of the island quarry. The gentleman he recommended was Colonel William Steuart of Baltimore.[13] Steuart was a perfect choice because his family still owned the one acre on the island. Steuart, Blagden, and Towson examined the quarry near the end of July. Steuart wrote back to Elgar, "I am of the opinion that there is every prospect of Mr. Towson being able to furnish the number required."[14] That is precisely what Towson did, for in March of 1825 Towson wrote to Elgar and stated that all column shafts would be in the district by May.[15]

Upon arrival at the district, a twenty-four-foot-long shaft[16] was given special care. Unlike smaller pieces of stone that could be placed on skids and pulled by oxen or placed in wagons and pulled by horses, each immense piece, "weighing eighteen tons," was loaded on a wagon and pulled by one hundred men. Anne Newport Royall wrote about transporting these shafts after her 1825 visit:

> They are brought from the wharf by the workmen, without the aid of horses, upon a strong carriage, made for the purpose - and a hundred men pull one with ease. This is quite a frolic for the men; and sometimes the members of congress will turn out in the evening to assist in pulling 'the big waggon,' as it is called, and join in all the pleasantry to which the novelty of the thing gives rise. When the column arrives at the capitol, it is cheered by loud huzzas from a hundred voices.[17]

Capitol's Completion and Quarries' Closure

Thomas T. Waterman
Historic American Buildings Survey/
Library of Congress

Author

Charles Bulfinch not only designed the Capitol's first dome but also gate posts and guard houses. In 1873, the above structures were removed from the Capitol grounds and replaced along Constitution Avenue.

Once shafts and blocks for the capitals arrived at the Capitol site, they were transformed by stone cutters. George Blagden was in charge of carvers in the sheds. The Italian stone carver Giovanni Andrei was, as Latrobe described him, the "Sculptor of decoration."[18] Just as he had done for the corn and tobacco capitals, he probably first created a plaster of Paris model prior to the workers carving the beautiful Corinthian capitals. One capital took carvers approximately six months to complete, and records indicate that $260 was paid for each.[19]

Tragically, George Blagden died on June 3, 1826, due to a cave-in on the West Front. The portico with its twenty-four stately columns was completed shortly thereafter, and construction on the Capitol was finally considered complete. (See chapter, "Extending the Capitol's East Front," to discover what became of the Capitol's columns.) Thirty-three intermittent years of construction had passed since the original cornerstone ceremony.[20]

Birthstone of the White House and Capitol

Forty sandstone columns in the crypt, which help support the Rotunda above, look today much like they did when Bulfinch placed them there in the early 1880s.
Architect of the Capitol

Even though the Capitol was considered complete, stone quarries were still operating in Stafford County in the 1820s and early '30s. Census records indicate that the highest percentage of laborers in Stafford County were those involved in quarrying occupations.[21] Census records do not list the slaves who worked in the quarries. They were employed, however, because in September of 1831 Stafford resident George M. Cooke wrote to Governor John Floyd of Virginia that after the Nat Turner rebellion, local slaves possessed a "high spirit of rebellion." Mr. Cooke was worried about the "over grown slave population," especially those concentrated at the stone quarries. He described the "defenceless conditions of the white population and the danger to which they are exposed."[22]

A finishing touch on the Capitol took place in 1829 when Bulfinch built terraces and steps on the west side, or present Mall side. Trees, flowers, and shrubs adorned the area. He also designed and added nine entrances to the grounds and two Gate Houses that were made of Aquia stone.

After thirteen years, Bulfinch left Washington in 1830. The position of Architect of the Capitol had been abolished the previous year.

Late in the 1830s, improved transportation made Washington, D.C., independent of local stone. Building materials arrived by ship from Europe, by rail from Baltimore, or by canal boat from above Washington. Changes in transportation and completion of many public buildings in Washington left the government quarries on the island dormant.[23]

Even though Government Island and other stone quarries were closed, they were examined in 1849 for possible use during the construction of the Smithsonian. The findings were published by Robert Dale Owen, chairman of the building committee of the Smithsonian Institution. His report stated:

> …public buildings erected by the General Government were built (with a single exception, the General Post Office) of a freestone so faulty, imperfect and perishable, that if it had been offered to the Committee charged with the construction of the Smithsonian Institution, delivered on the ground for nothing, they would, even then, have rejected it…. It cost the Government, delivered to the city, never less than forty-five cents, often fifty, per cubic foot of dimension stone.[24]

Taken in 1860, this photograph shows Thomas Walter's unfinished cast iron dome. In the foreground is the Washington Canal, which was later filled in.
Architect of the Capitol

The Smithsonian instead used a strong red sandstone from Seneca Quarries in Maryland. Aquia stone was also not considered for the 1850s Capitol expansion.

In the 1850s, the size of the fifty-year-old Capitol seemed inadequate. Increases in the membership of Congress, along with an increase in tourists, required an addition. In September 1850, $100,000 was appropriated to add two new wings.[25] These wings dwarfed the original sandstone wings and were constructed of white Tennessee marble. Improved transportation, including the Chesapeake and Ohio Canal, made possible the extensions on both sides. The new marble walls, created of a stronger stone, would not deteriorate as much as freestone and would not need constant painting. (Today, the original sandstone wings appear as links to the larger marble wings.)

A design competition, authorized by Congress, was held to seek a designer. Thomas U. Walter, a mason and architect, submitted the winning entry. Walter was appointed in 1851 and became the fourth Architect of the Capitol.[26]

On July 4, 1851, the Capitol had its third cornerstone laying. This ceremony celebrated the Capitol's enlargement. *The National Intelligencer* reported, "The day was ushered in by salutes of artillery from different points of the city, and as the glorious sun gilded our tallest spires, and shed a luster on the dome of the Capitol, it was welcomed by a display of National Flags and the ringing of bells from the various churches and engine houses."[27]

During that year, a fire gutted the Library of Congress, which was housed in the Capitol. Priceless volumes were destroyed by the fire that reached through the windows on the west side and damaged some original stone.[28] This was the second time that fire destroyed the library, so Walter decided to replace the interior walls with cast iron, making the library fireproof. Walter became fascinated with cast iron and discovered its use for creating elaborate designs such as rosettes and scrolls on walls and ceilings.[29]

Before the library was completed, it became apparent that the Bulfinch dome needed replacement. It leaked and looked strangely small with the new marble extensions.[30] In 1855 Congress authorized a replacement. Walter, who had studied in Europe, recalled observing various domed structures such as St. Paul's in London and St. Peter's in Rome.[31] These influenced his new dome plan. Walter's success with cast iron in the Library of Congress caused him to select this heavy material. Plans were drawn for inner and outer cast iron shells, weighing almost 9 million pounds.

Would the original walls be strong enough to support the new addition? An engineering report stated that "the foundation walls are formed of large bluestone, laid in lime mortar. The basement walls are of Aquia Creek cut sandstone. The principal story walls, or the walls of the rotunda, as high as the interior cornice, are faced on the inside with Aquia Creek sandstone, and are backed with brick."[32]

It was calculated that the dome would exert a pressure of 10,000 pounds per square foot on the rotunda walls.[33] A hydraulic proving machine was used to test the crushing weights. The findings per square foot were as follows:

Aquia creek sandstone	755,280 lbs.
Hardest Brick	1,849,248 lbs.
Brick and mortar, two years old	339,120 lbs.
Brick	307,277 lbs.

An engineering report concluded, "Now, the greatest pressure which will be exerted by the new dome is at the basement, or crypt floor, where it will be 13,477 pounds per square foot. The stone which is to bear this weight requires a pressure of 755,280 pounds per square foot to crush it, or about fifty-six times the weight of the dome."[34] The findings determined that the Aquia freestone, selected some sixty-five years earlier, could support the weight.

While discussions were going on concerning the strength of the stone, there was a little activity on the closed Government Island. Thomas P. Towson, Jr., who was then property owner of the land adjoining the island, complained through Congressman William Smith about squatters on the abandoned quarry. He said a dry connection at low tide between his land and the island made it a "port for stragglers." Towson was told that the government land was regarded as nearly exhausted of stone and of "inconsiderable" value. The Commissioner of Public Buildings thought the sale of land by an act of Congress would be "tedious and hard to get at," so the Secretary of the Interior authorized the appointment of Towson as "agent" to take charge of the island. This, the easiest way to solve the problem, let Towson protect his own interests.[35]

Meanwhile, in the district, objections were raised as to the feasiblity of constructing a cast iron dome with money and material needed to fight the Civil War. President Lincoln intervened, saying that construction should continue and a crowning statue should be raised "as a sign that the Union is going on."[36] In December 1863 the construction was sufficiently advanced to allow the new dome, resting upon Aquia stone, to be topped with the Statue of Freedom—symbolic in the year of Union victories at Vicksburg and Gettysburg.

The island remained inactive even during the Civil War. Only 800 yards away on the mainland, war relics have been found. Six miles away, near the mouth of Aquia Creek, the first engagement between Northern naval vessels and Southern shore batteries occurred. The Confederacy defended Stafford County until the spring of 1862, when

Today one can see the markings made by Confederate and Union forces that occupied Aquia Church during the Civil War. Names, regiments, and dates were scratched in the quoins and appear as fresh as the day they were originally etched.
Lou Cordero, Fredericksburg Free Lance-Star

they moved south. Except for brief intervals, the Union controlled the Stafford area. Union soldiers used the area for their winter quarters. One hundred thirty-five thousand men camped between Falmouth, on the Rappahannock River, and Aquia and foraged the county until nothing was left.[37]

Margaret Waller Ford, who lived at Woodstock after the Brents, wrote a letter describing the unfeeling Union soldiers. She wrote, "The federal forces carried all the [freestone] tombstones into camp to make fireplaces in their tents that could be moved. Some of the Brent stones have inscriptions on silver plate set in the stone, - these the Yankees picked out and sent north."[38]

Union troops occupied quarrier William Robertson's house that Latrobe visited sixty years before, as well as a stone house that was attached to it.[39] Nearby at Aquia Church, Union cavalry stabled horses in the pew boxes.[40]

A local historian, John T. Goolrick, wrote, "When the war ended Stafford was utterly devoid of stock, food and forage, and the soil had gone down or grown up into brush."[41]

Capitol's Completion and Quarries' Closure

The people of Stafford County had to eke out an existence without horses, money, or slaves. They were not concerned with reopening quarries or preserving their history. They were just concerned with restoring their lives.

Around five decades later, in the early 1920s, one quarry reopened. Rock Rimmon, or Rock Raymond, which sat upon a bluff about two miles from Government Island, was purchased by the George Washington Stone Corporation. Located on Aquia Creek, the quarry had previously contributed stone for the Capitol. The corporation's literature stated that the quarry was "covered with an almost impenetrable forest," but they cleared the area and built miles of roads in order to get equipment into place. A brochure for their stone said that "ox-carts have been replaced by the railroad and the channels are cut only four inches wide by modern self-propelled channeling machines. The quarries, in fact, are equipped with the most modern machinery obtainable."[42] The corporation also acquired a stone cutting mill in Alexandria and was able to sell tons of quarried and cut freestone to a new market.* It advertised that its stone could be shipped from Jacksonville to New York. This quarry closed after 1931, when the corporation's Alexandria cutting mill burned.

This photograph shows a channeling machine used at the George Washington Stone Quarry from the turn of the century until around 1931. *H. P. Caemmerer*

*See chapter "Other Structures" for list of structures that used George Washington Stone Corporation's Aquia stone.

(Aquia Overlook, a residential community, now sits on the quarry site.)

**The picks and mauls are silent now at Government Island.
Gone are the muscled laborers who quarried the rock.
Gone, too, are the masted ships that carried it north.
Government Island is quiet now**
    Overgrown with oaks and maples and sticker bushes.
Only the silent stones say that
    This was once the nation's most famous quarry.
            Jim Hall, *Fredericksburg Free Lance-Star*, July 6, 1992

18

Truman's White House Renovation

In 1948, President Harry Truman complained to the Commission of Grounds and Buildings about the White House's condition. He had heard noises in the night and was convinced the floors were fragile and even dangerous. He wrote in his diary about the state of the floors. "Fields, the head butler, brought my lunch to my desk in the Study. Fields is a big man—and a grand man too. The floors sagged and moved like a ship at sea."[1] His daughter, Margaret Truman, wrote in her book, *Souvenir*, that "…the need to repair it became dramatically evident…when Dad went upstairs and found that my Steinway piano had fallen through the floor. A large hole had opened up in the second floor and the piano was sitting there tipsily, with one leg in the hole."[2] President Truman told his sister about Margaret's piano and added, "Now my bathroom is about to fall into the Red Parlor. They won't let me sleep in my bedroom or use the bath."[3]

After preliminary inspection, the President was told the second floor was in danger of coming down. Also the new roof of 1927 and the third floor put pressure upon interior brick walls and piers, producing cracking when settling. Wood throughout the building had dried. Even plaster on the East Room ceiling was sagging.[4] An engineer told Truman that the State Dining Room ceiling remained in place from "force of habit."[5] After thorough examination, an investigative committee wrote, "The building violates principles of good fire engineering practice

Birthstone of the White House and Capitol

Debris chutes enabled workers to gut out the interior.
Abbie Rowe, National Park Service, The White House

and presents a definite fire hazard to persons and property."[6] Thus, a renovation project was begun, tours terminated, and the White House evacuated. In November of 1948, in time for Thanksgiving, the Trumans moved into the Blair House across the street.

There was considerable debate in Congress as to the method of reconstruction. Some believed that the entire building should be demolished and the walls should be reproduced in limestone, granite, or marble. Others thought the old sandstone walls should be removed piece by piece and labeled, and the inside should be rebuilt with concrete and steel with the stones placed back in proper position.[7] Harry Truman had other plans. He wrote in a May 1949 letter to Congressman Clarence Cannon:

> My suggestion is that we do not tear down the present building. The outside walls are in good condition and are on a fairly good foundation. The difficulty is with the supporting columns inside of the White House, which are not on a firm foundation - they are made of bricks and the bricks themselves are disintegrating. I believe, with the right sort of a contractor and with proper supervision, the interior can be removed and suitable foundations put under the supporting pillars. We could put a steel and concrete structure inside the walls and restore the inside of the house to its original condition. We are saving all the doors, mantles, mirrors, and things of that sort so that they will go back just as they were before.[8]

A renovation commission was formed and considered plans. One member of the commission, Louis C. Rabaut, a congressman from Michigan, said that if the interior was to be removed, the stone walls should, too. However, the commission voted for Truman's plan, with only Rabaut casting the dissenting vote.[9] Thus, it was President Truman's interest in architecture and his desire to preserve history that kept the Aquia sandstone walls from tumbling down.

The first step in reconstruction began on December 7, 1949, with the underpinning of the exterior walls. Pits were dug under the freestone walls to a depth of twenty-four to twenty-seven feet. Then concrete was poured, creating footings that rested on a firm stratum of sand and gravel. Previously the walls had rested upon clay.[10]

The day after the underpinning work commenced, dismantling was started. Interior pieces such as ceiling ornamentation, windows, doors, chandeliers, door trim, hardware

Shown below is a myriad of activities during the Truman renovation.
Abbie Rowe, National Park Service, The White House

White House bricks were given to Mount Vernon to reconstruct George Washington's green house. *Author*

and mantlepieces were photographed, numbered, and sent to storage. Some original furnishings were loaned to the Blair House, while others were placed at the National Gallery for safekeeping. Some surplus mantles were given to museums. Bricks made in kilns on the White House grounds in the 1790s were shipped to Mount Vernon. Nearly 95,000 bricks were used for both reconstructing George Washington's orangery, or greenhouse, and restoring garden walls. Mount Vernon also received ten Army truckloads of large blocks of Aquia stone. They were dumped in a storage yard President Washington called *Hell Hole*.[11] Over 10,000 bricks and 4,000 square feet of flooring were used at Fort Myer, an Army post in Virginia. Fort Belvoir, another Virginia Army post, received pine doors, scrap lumber, brick, and plaster.

Citizens were also afforded an opportunity to purchase a piece of the White House. Pieces of Aquia stone, brick, old square nails, copper wire, and sections of pine were sold in kits ranging from 50 cents to $2. One could even buy enough brick to construct a fireplace for $100.[12]

Dismantling the interior took approximately three months. The tedious process of saving as much as possible completed, all remaining interior debris was shipped to Fort Myer and used there and elsewhere as landfill. The debris was sent out via shafts and window chutes and carried away in dump trucks. By June 1950, the demolition was 96 percent completed, and underpinning was 98 percent completed. In all, 126 pits were dug and piers placed.[13]

The Aquia sandstone walls now formed an empty shell. Sections of bare stone were exposed, while other sections had portions of the old, reddish-orange brick clinging to interior rock faces. Now it was time to shore up the walls with structural steel. Bethlehem Steel Company was awarded the contract for the interior steel frame that would support all interior loads.

A two-story basement was dug. President Truman did not want any sandstone wall touched, so all the equipment, including bulldozers, had to be disassembled and inserted through the window holes and reassembled in the interior. After the basement was dug below the ground level of the freestone walls, an exit was created for the heavy equipment and trucks.

President Truman had an abiding interest in both history and architecture. He cared deeply about every detail of rebuilding

According to Truman's instructions, vehicles were dismantled and placed within the White House shell during restoration. They were then reassembled, thus protecting the Aquia freestone walls. A new basement level was dug out as well as access to the outside. An interior steel frame and underpinning of the walls helped ensure the stability of the renovated White House.
Abbie Rowe, National Park Service, The White House

Birthstone of the White House and Capitol

Mason marks were uncovered during Truman's White House renovation. Some stones were removed and sent to Masonic Lodges in North America. More than forty marks were discovered in the White House alone. Mason marks were also discovered at the Capitol.
White House Historical Association/White House Collection

and preserving this historic structure. Rex Scouten, former White House Curator, was a Secret Service agent during the White House renovation. He recalled walking over to the site on numerous occasions with Truman. "When the walls were propped up during construction, he would crawl up and walk on the catwalk to inspect the area."[14] Margaret Truman wrote, "Dad went there every Sunday, just like any other householder, to note the progress or the lack of it and then to prod the people in charge, again like any householder. The repairs went on interminably, and sometimes it seemed to me that the contractors must have thought the Democrats had had the White House long enough!"[15]

During renovation, many stones with mason's marks were found. Just as an artist signs his name upon his finished masterpiece, a stone mason would carve his special symbol upon a stone and leave it within the walls of the house or structure on which he had worked. The marks were usually geometric in design. More than forty different marks were found in the White House alone. It has been interesting for scholars to see which artisans worked on both the President's House and the Capitol. By comparing identifying marks, it is even possible to trace a mason's roots earlier to structures in Europe.

President Truman set some of the stones with mason's marks in the kitchen on the ground floor, today's curator's office. He sent others to Masonic Grand Lodges in all forty-eight states along with a letter. A portion of that letter stated:

> I place in your hands…one of the…stones removed from the walls of the White House during its restoration and rebuilding. …These evidences of the number of members of the Craft who built the President's official residence so intimately aligns Freemasonry with the formation and the founding of our Government that I believe your Grand Lodge will cherish this link between the Fraternity and the Government of the Nation, of which the White House is a symbol.[16]

The renovation project, intended for completion in a year, actually took three years and four months. Now the Aquia stone walls encased a steel skeleton with concrete partitions and floors. In his diary Truman wrote, "They…put in steel and concrete like you've never seen in the Empire State Bldg., Pentagon or anywhere else. Only an earthquake or an atomic bomb…could wreck the old building now."[17] Wood and masonry floors were laid, and plaster and paneling adorned walls. Original levels were restored, while the two new levels below ground housed such things as electrical equipment, plumbing, and even a bomb shelter. Most old furniture was renovated and returned. New furnishings were purchased. For example, new curtains graced the windows while silk damask coverings were applied to some walls.[18]

On March 27, 1952, after a vacation in Key West, Florida, Truman returned to the newly renovated White House. After examining the refurbished mansion, he recorded in his diary, "I spent the evening going over the house. With all the trouble and worry, it is worth it—but not 5½ million dollars! If I could have had charge of the construction it would have been done for half the money and in half the time!"[19] The total cost was $5,832,000.

About a month later, Truman took representatives of the major television networks, CBS, ABC, and NBC, on a guided tour. Thirty million viewers turned in and were captivated with the newly restored "People's House." Truman did not use a text but displayed his knowledge of both architecture and history to an enthralled audience.[20]

Lorenzo Winslow, White House Architect, wrote before the restoration, "It is the President's desire that this restoration be made so thoroughly complete that the structural condition and all principal and fixed architectural finishes will be permanent for many generations to come."[21] And that is exactly what Truman accomplished. His vision to see the original exterior preserved had been executed and, thus, the Aquia stone walls are the only remaining portion of the original structure standing today.

19

Extending the Capitol's East Front

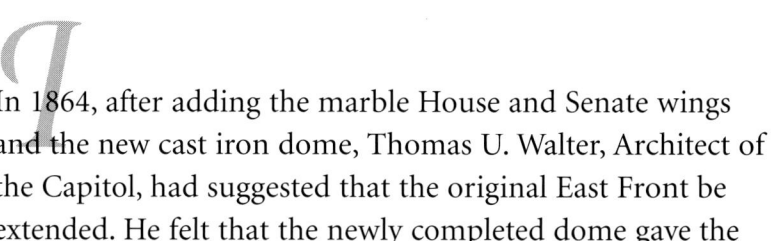

In 1864, after adding the marble House and Senate wings and the new cast iron dome, Thomas U. Walter, Architect of the Capitol, had suggested that the original East Front be extended. He felt that the newly completed dome gave the impression that it was hanging over the central portico and was not supported adequately.[1]

Approximately nine decades later, in 1952, President Truman expressed the same opinion. *The New York Times* reported that Truman said, "…the Capitol had not been finished according to the architect's plan—'the iron dome' is not on it right."[2]

Several years later, J. George Stewart, Architect of the Capitol, suggested Walter's extension again. The extension would add needed office space as well as improve the Capitol's appearance. Many objections surfaced as few wished to change the original building. The Daughters of the American Revolution were particularly opposed. President Truman expressed his opinion again to a friend in a letter dated February 15, 1958:

> …If you will go out and stand on the House steps or Senate steps…you will find that the dome is set seven or eight feet over the front portico of the Capitol. I am 100% for the completion of the center front.… If these people who are crying about the Capitol now really think they are doing the right thing, they ought to continue their efforts by taking the present dome off and putting the old one back.[3]

Birthstone of the White House and Capitol

The East Front extension was started on July 4, 1959, and completed in time for John F. Kennedy's inauguration on January 20, 1961. Every detail of the Aquia stone façade was duplicated in Georgia marble. *Architect of the Capitol*

In 1956 and in 1958, the proposal continued in debate, but Congress finally voted in favor of the extension. J. George Stewart was ultimately responsible for the remodeling, but it was Sam Rayburn, House member for forty-eight years, who was instrumental in getting the bills through Congress.[4]

A cornerstone was laid for the extension on July 4, 1959, by President Dwight D. Eisenhower. A traditional Masonic ceremony followed. Most of the Aquia stone walls were left standing. A marble extension was built thirty-two feet six inches east of the sandstone walls. The twenty-four Aquia stone columns and decorative portico pieces were removed and placed in a government storage facility, the Poplar Point Nursery, by the Anacostia River.

Every effort was undertaken to faithfully reproduce the original Aquia stone facade out of white marble from Georgia. No longer would the columns and facade need to be painted white as before. The original freestone exterior walls became interior walls. The extension created ninety much needed new rooms. East portico construction was completed for President John F. Kennedy's 1961 inauguration.

Extending the Capitol's East Front

For almost three decades, the columns that Towson had worked so hard extracting from the island lay like outcasts in the government's storage depository. Mrs. Ethel Shields Garrett, a philanthropist, led a fight to move the historic columns to the National Arboretum, where they could be enjoyed by the public. Russell Page, a noted English landscape architect, was asked by Mrs. Garrett to design a plan to display twenty-two of the original twenty-four sandstone columns. Unfortunately, neither Mr. Page nor Mrs. Garrett lived to see the columns set on a knoll overlooking a meadow in June of 1990. Now columns surround a lovely reflecting pool and flowing fountain. Steppingstones on the green grass were once Aquia stone facing stones from the East Front. The job was costly, but $1.5 million was raised from private contributions.[6] The task of moving the stone was difficult, as each twenty-four-foot shaft weighed twenty tons.[7] Architect Reed Black, who worked with the firm of Oehrlein and Associates in assembling the columns, found the Aquia sandstone to be "a better quality stone with few inclusions."[8] Today the lovely Corinthian columns stand majestically on the hillside where they can be viewed by thousands of people annually.

Top: The Aquia stone columns from the original Capitol were placed in storage for around thirty years. They were later removed and placed on a hill at the National Arboretum in 1989.
Oehrlein and Associates

Bottom: Twenty-two of the original twenty-four freestone Corinthian columns that once created the East Front Portico now stand regally at the National Arboretum amidst a fountain, water stair, and reflection pool.
© 1992 Southern Living, Inc. Reprinted with permission

* Stone from the East Front extension was used to replace damaged stone at the White House in the 1990s. See chapter, "Painting and Restoring the White House Stone," for more details.

20

Who Owned Government Island?

Although legally the property of the federal government, Government Island was considered unclaimed property after the Civil War. Since it had been abandoned, the state of Virginia mistakenly took control of the island. Therefore, the Governor of Virginia, Frederick W. M. Holliday, conveyed the island to Samuel B. Howell and C. A. Bryan on April 1, 1879. After that, the property was sold or willed repeatedly with all owners blissfully unaware that their island actually belonged to the United States.[1]

In the early 1960s a unique turn of events finally unearthed the true ownership of Government Island. Surprisingly, it took nearly nine decades for the truth to surface. Lawyers Title Insurance Company of Richmond was asked to search title for new owners of 1,600 acres of land including an island, "Wiggington's Island." Their research attorneys showed the island's title going back to 1878 from the Governor of Virginia. Since only sixty years of clear title were needed, the policy was issued.

The newly insured became disturbed when they found signs on the island reading "Keep Off—Government Property." Lawyers Title called Washington, D.C., but government agencies said that they knew of no government-owned land in the

Birthstone of the White House and Capitol

> **Lawyers Title settled the claim with their insured by paying them the value of the island.**

area. The owners tore down the signs upon discovering they were posted by a history buff.

A few years later, the insured's attorney contacted Lawyers Title. The owners were disturbed to read a newspaper article stating that "Government Island" or "Wiggington's Island" was to be sold at public auction by the General Services Administration.

Lawyers Title was very concerned to hear such news and was shocked when it received copies of the instruments by which the United States claimed title. They realized then they were dealing with "sacred soil" dating from 1694. They also discovered that one acre (Robert Steuart's—1786) did not belong to the government. Lawyers Title settled the claim with their insured by paying them the value of the island.[2]

Government Island was auctioned in 1963. The Stafford Board of Education wished to purchase the island for an outdoor classroom, while the Commonwealth of Virginia was thinking about using it as a state park.[3] However, when time came for interested parties to submit their bids, there were only two entries. Thomas Metts, a Stafford County resident, offered to purchase the island for $6,000 and sealed his bid in an envelope. He had frequently camped on the island with Boy Scouts and wanted it for their use.[4] Farrar A. Simons, who worked at the Government Accounting Office, had the winning bid of $6,345.63, just $345.63 over Metts' bid. Simons and his wife planned to build a retirement home on the isle, and he was pictured riding atop a tractor in a newspaper article.[5] Not disturbing any of the quarry sites, Simons bulldozed a very small portion of the island. Their little retreat no longer stands on the grounds. No one ever built on the island again.

21

Deterioration of Aquia Stone

During the War of 1812, when the British burned the Capitol and White House, Aquia Creek freestone was heated and then cooled immediately due to a sudden rainstorm. The result of these violent physical events was damage to the stone. This required replacement and reconstruction on both structures.

In addition to the extremes in temperature that occurred in 1814, there were, and are, other forces, both natural and man-made, that can cause stone deterioration:

- All stone, not just sandstone, is porous. It absorbs and retains moisture. Moisture can be from rain, water vapor in the atmosphere, groundwater, and even condensation from the interior of the building. Water within a stone can usually rise to the surface and evaporate. If this does not occur and water remains trapped in the stone, external temperature conditions, such as freezing and then thawing, can cause damage.

- Soluble salts can sometimes be found within stones. Salts, too, can rise by capillary action from the ground and lodge in the stone. If these salts crystallize in the pores of the stone, they can cause surface breaking or give an efflorescent appearance.

- In metropolitan areas, carbon dioxide and sulfuric and nitrous oxides can be dissolved in rainwater. This solution can create disintegration. Also, if the stone itself is dirty, it can attract more moisture.[1]

- Mortar joints often decay, and modern cement is used in its place. Since Portland cement mortar, for example, is much harder than the lime mortar used in the original construction, it will actually crush sandstone when a freeze/thaw cycle occurs. Cement expands during temperature changes and can even pulverize weaker stone. Unfortunately, restoring stone structures by filling in missing chunks of stone with cement creates more disintegration than would normally occur.

- Sealants can cause damage. For example, Thompson's Water Seal was applied to Aquia stone at Old Cape Henry Lighthouse and was believed to be a source of some exfoliation, or flaking.[2]

Aquia stone's capacity for resisting deterioration and its merit as a building stone have been questioned throughout the years. For example, it was evaluated for use in the Smithsonian in 1849. In 1856, it was tested to see if it could withstand the weight of the Capitol's cast iron dome. Debates about preserving the stone arose in Congress during the late 1950s when the Capitol's extension of the East Front was discussed. Even today, controversy surrounds the stone.

In 1991 the Association for the Preservation of Virginia Antiquities (APVA) contracted with architects to evaluate the condition of the Aquia Creek sandstone used in the Old Cape Henry Lighthouse at Fort Story, Virginia Beach. Most of the exterior of the shaft is Rappahannock red sandstone, but Aquia Creek sandstone can be found in the foundation as well as trim.

> The architectural firm of Wood, Sweet and Swofford was hired to evaluate some previous suggestions to stop deterioration of the Aquia stone and make their own recommendations.

The architectural firm of Wood, Sweet and Swofford was hired to evaluate some previous suggestions to stop deterioration of the Aquia stone and make their own recommendations. One suggestion was to patch the eroded sandstone with a type of cement. Another recommended that the eroded stone be removed and replaced with a cast stone made from an Aquia Creek stone aggregate. Another strategy was to chemically treat the stone with a strengthener. Considering that there is no such strengthener, that an aggregate stone would be incompatible with the existing stone, and that a cement mix has been known to delaminate (separate into layers) the sandstone, another alternative method was suggested. The firm believed that the deteriorating stone should be replaced with Aquia Creek stone,[3] which had been used in 1770 when the lighthouse was constructed. Some 7,000 tons of stone had been shipped to the site. Calculations indicated that only 1,600 tons of stone were used by the builder. The remaining 5,400 tons remained below the sand.[4] Since only six to eight stones needed to be replaced, they concluded that, "Surely such blocks could be salvaged from the great amount of stone scattered about the lighthouse site."[5]

In their report, the architects stated that historical architects had considered Aquia sandstone to be "an unsound building material." They concluded, however, "The two hundred years it has existed in the Lighthouse suggests the contrary."[6]

The merits of Aquia stone were also discussed at a conference for the Bicentennial of the Laying of the Capitol's Cornerstone on September 18, 1993. William C. Allen, Architectural Historian for the Office of the Architect of the Capitol, addressed the conference. He commented that Aquia stone was an "inferior stone."[7] Later, when questioned, he clarified his statement by stating that it was "inferior" since so much of it deteriorated on the east side of the Capitol, necessitating replacement with marble. Also, he mentioned that the stone on the west side of the Capitol had to be replaced, and only 60 percent of the original Aquia stone remains.[8]

A different view is held by Rex Scouten, former White House Curator, who believes the Aquia Creek stone to be of good quality, as it has held up quite well throughout the 200 years.[9] The difference in the withstanding properties of the stone on the two buildings may be due to the fact that the White House was whitewashed over a decade before the Capitol, perhaps preserving its quality. Controversy surrounding the stone will probably continue as long as both structures remain standing.

22

Painting and Restoring the Stone of the White House

Most elementary school textbooks state that the White House was given its name after the building was painted white to cover scorch marks from the British burning in 1814. Actually, the house was painted sixteen years before the burning with whitewash used to protect the stone's porous surface. Hoban wrote in the fall of 1798 that the workmen were engaged "in cleaning down and painting the wall of the building and striking the scaffolds."[1] His records show that three men were paid for nineteen days' labor painting the President's House.[2]

The formula for this 1798 wash was a mixture of lime, rice, glue, salt, and Spanish whiting (a white, powdery substance created by grinding a calcium carbonate substance such as chalk). The following is the original recipe:

> Take half a bushel of good unslacked lime, slack it with boiling water, covering it during the process to keep in the steam. Strain the liquor through a fine sieve or strainer, and add to it a peck of clean salt, previously dissolved in warm water; three pounds of good rice ground to a thin paste and stirred while boiling hot; half a pound of powdered Spanish whiting, and a pound of clean glue, which has been previously dissolved by first soaking it well, and then hanging it over a slow fire in a small kettle, within a large one filled with water. Add five gallons of hot water to the mixture; stir it well, and let it stand a few days, covered from dirt.[3]

Since most contemporary buildings were brick, this new white structure stood out, creating the name "White House." In 1810, the *Baltimore Whig* first used the term "White House" in an article.[4] Evidence of common usage is reflected in Massachusetts Congressman Abijah Bigelow's letter to his wife. On March 18, 1812, he wrote, "There is much trouble at the white house as we call it, I mean the Presidents.'"[5] In 1902 President Theodore Roosevelt made the name official by adding it to his personal stationery.[6]

When the White House was painted after the British burning, the paint formula was changed. Linseed oil and clay were used to make a stronger, thicker mixture to insure covering the fire damage on the charred walls.[7]

After the White House was rebuilt (1814–18), it was painted periodically. But in the 1890s, a policy was established for painting every four years in order to spruce up the building for a newly elected President. Money for this task was appropriated every four years by Congress. In 1976, something unusual occurred. The paint would not adhere. Rex Scouten, Head Usher of the White House during that time, recalled the details:

> …'76 was a painting year and we were aware that new coats weren't adhering and old coats were coming off in sheets in certain areas. The big sheets came off in one foot by two foot squares. We realized something had to be done, so we started a study at that time. We put the team together and first realized, obviously, that we were dealing with sandstone and it was too soft to sand blast. We'd have to find some other method of removing the paint and removing it in such a way where we wouldn't be destroying the beautiful carving on the house.[8]

At Mr. Scouten's request, a team was formed that consisted of a representative from the Architect of the Capitol's office and members of the National Park Service [who have managed the White House and grounds since 1935] and the National Bureau of Standards [currently known as the National Institute of Standards and Technology]. Rex Scouten actually initiated the team and the paint removal search, but he credits Elmer Atkins as the person who did the leg work. At that time Mr. Atkins was the National Park Service Assistant Regional Director for the White House Liaison.

Four contractors were contacted and each given an exterior area at the northeast corner at the basement level of the White House to demonstrate their method of removing paint. The four methods tested were:

- Chemical treatment with a high-pressure water spray;
- Hand scraping and chipping;
- Abrasive blasting with aluminum beads;
- A combination of hand cleaning, chemical treatment, and water spray.

The chemical treatment with the high-pressure water spray was the most effective. A Connecticut-based firm, RUDCO (now RAMCO), received the contract. Previously, the firm had successfully cleaned the freestone in the Capitol's interior.[9]

A Washington, D.C.-based company, Duron Paints, developed paint specifically for sandstone. The paint coated and protected the stone yet was one that could be used consistently without much buildup.

It took several years to form the team, select the best paint removal method, and develop a paint. After an oil alkyd paint was developed by Duron, Rex Scouten went to President Jimmy Carter and told him about the paint situation. Mr. Scouten recalled, "All the [appropriated] funds I had were for the repainting. If I could make do, would he [the President] be agreeable to having me remove the paint from the east side? He gave his approval."[10] Mr. Scouten then went to RUDCO and asked them to remove the paint from the east elevation and paint the entire house within the appropriation limits. Dr. Feinburg, the president of Duron Paints, generously donated the paint. So both tasks, removing paint on the east facade and painting, were accomplished using the money appropriated only for one.

Protective coverings and scaffolding were used during the paint removal process at the White House. The South Front is shown in this 1991 photograph.
Jack Boucher, Historic American Buildings Survey, The White House

When asked about his role in the process, President Jimmy Carter wrote, "I am glad to have authorized the removal of paint on the exterior of the White House. This has revealed some of the original wonders of the building construction."[11] And that it did, for in some places the paint was so thick that it filled in details of delicate roses, graceful leaves, or flowing ribbons carved of Aquia stone. Some areas had as many as forty-two layers of paint with a thickness of one-quarter to three-quarters of an inch.

RUDCO removed the layers by painting on various chemicals. Chemical formulations had to be changed three times to remove all the layers correctly without damaging the stone. One solvent worked well with paint from one time period but would not work well with paint from a different era. Old paint was scraped off with a plastic hand tool to reduce damage. A hot water solution of mild acetic acid was applied at approximately four to six gallons per minute. This balanced the alkalinity of the chemicals, achieving a neutral pH.

Some areas, especially over the North Door, had intricate carving and thick layers of paint and suffered deterioration from repeated chemical treatments. Another approach was needed. Heat guns were used on the outer layers to remove the bulk of the coatings. Later, a chemical remover was applied.[12]

Historic American Buildings Survey documented the White House for the 200th anniversary of the laying of the cornerstone. Photographs and drawings were made both inside and out. The drawing shows the north side. The iron oxide streaks of each individual stone were also recorded.
Library of Congress

Paint layers on the White House walls consisted of all types: whitewash and clay-linseed oil; modern oil-based, latex, and lead-based formulas. Since some layers were lead-based, the Environmental Protection Agency ordered all waste to be collected and disposed of properly. Paint and rinse water was collected with wet vacuums and stored in fifty-five-gallon drums for later disposal.

The linseed oil and clay layer, applied after the British burning, was the most difficult to remove. Mr. Scouten said the original whitewash was never found. "The other paint that they put on after the fire was so heavy that it drew it right up. We tried all outside areas, but we have never been able to find it. It is not that unusual because it was a much thinner coat."[13]

By October 24, 1980, all east wall paint was removed. In 1981 it was decided to strip and repaint all remaining walls. The paint removal project went on four-year cycles and lasted approximately eleven years. The last section was completed in 1991.

Removal of the many layers of paint uncovered rich stone carving at the White House. None is more evident than that over the entrance at the North Portico. The quality of the carving and its beauty with intricate details reveal superb craftsmanship.
Jack Boucher, Historic American Buildings Survey

Before sections of the White House were repainted, Historic American Buildings Survey and National Park Service, documented the White House. This was completed for the 200th anniversary of the cornerstone laying. Photographs and drawings of the stone were made. The drawings were so detailed that they showed the veins or reddish ferrous oxide markings on each individual building block.

Robert Kapsch, former Historic American Buildings Survey Chief, recalled first seeing the photographs taken by Jack E. Boucher of the north entrance carvings. "The sculpture that had been obscured by layer upon layer of paint was probably the finest building sculpture executed in the United States in the eighteenth century."[14]

Birthstone of the White House and Capitol

James I. McDaniel, former U.S. Parks Liaison Officer, was amazed to see the color of the naked freestone on the east wall. He recorded his thoughts:

> The sand-colored stone had an almost rose cast when dry. But when rained upon, the sandstone absorbed considerable water and turned a dull gray. When the sun reappeared after a rain, it was possible to watch the water drain slowly down through the stone, as if the house were a wet sponge set on its end. The reasons for the original whitewash sealer, as well as the subsequent paintings, became quite clear during the observations of that winter.[15]

Exposed walls revealed damage to the freestone. The problems were:

- CRACKS—Hairline cracks in the original stone expanded by years of moisture penetration.
- EROSION and EXFOLIATION, or flaking—Due to weathering.
- MISSING STONE—Sheet metal had been molded to cover missing or eroded dentils and modillions and then painted to duplicate the appearance of stone.
- SPALLING, or peeling away of the freestone's surface.
- CEMENT PATCHES—Unbeknownst to the repairmen in the nineteenth century, cement was harder than the stone, causing the stone to be overstressed and fail to bond with the cement.
- RUSTING of IRON BOLTS and CLAMPS used by early craftsmen. The rusting caused an expansive force, putting pressure on the stone.[16]

Once these problems were uncovered, it was decided to launch a stone renovation project. In 1984 Vincent Palumbo, an Italian stone carver who previously worked at the National Cathedral, was selected for stone restoration on the North Portico columns. In 1987 he and Stone Carving and Restoration Enterprises, Inc., of Washington received a National Park Service contract. Palumbo said the building was solid and damage was minimal but discovered that not to be the case. Some metal boxes that covered cracked dentils retained moisture. Palumbo stated, "I took off one of those boxes and the stone was just like mud, like mud."[17]

Paint covered up damage created by the environment as well as repairs that were accomplished with cement and metal inserts.
Bill Allman, The White House

Painting and Restoring the Stone of the White House

By 1990 there were fourteen workmen on this project. During this year Patrick Plunkett, master stone carver and mason, was chosen to be the new stone superintendent. Plunkett, a Scotsman like the original masons, had previously worked at the National Cathedral. When he took over, he reduced the work crew to four experienced men besides himself.[18]

It was decided that rather than remove huge blocks of stone, only damaged portions would be removed. Lee H. Nelson, author of *White House Stone Carving*, described the work as being "more like dental fillings than new teeth."[19] Large blocks of stone were replaced, but whenever possible, only small, damaged sections were removed.

Since the Aquia quarries had been closed for decades, and it would have been impractical to reopen them, the question was raised if it would be possible to use the discarded Aquia stone from the Capitol's East Front extension that had been placed in government storage facilities since 1959.

White House Chief Usher Gary J. Walters worked with George M. White, Architect of the Capitol, to obtain Congressional approval for use of the Capitol's discarded stone. George H. W. Bush, then Vice President, was instrumental in obtaining the stone. Later, President Bush recalled his efforts to use the stone and wrote to Chief Usher Walters after receiving information about the project, "Thanks for the brochure. Yes, I did sign the stone letter—gladly."[20]

The freestone that was removed from the Capitol's East Front extension in the late 1950s was placed in government storage. Some pieces were used in the White House's stone restoration project. Patrick Plunkett, master stone carver and mason, tried to match the Capitol's discarded stone with stone needed for repair at the White House.
Tim Buehner, National Park Service

Birthstone of the White House and Capitol

During the 1990s stone superintendent Plunkett visited the storage facility and tried to match the White House's damaged stone with the Capitol's discarded stone. He attempted to find Aquia stone with approximately the same color, markings, and texture as the old defective stone. Therefore, the remaining stone and the replacement stone in the original portion of the White House were quarried in Aquia Creek over two centuries ago.[21]

The salvaged stone was transported to the White House and placed by or in the stone-carvers' workshop, located on the south side of the building. Templates were made, and the stone was cut so that it would fit snugly into an exact location. However, the face of the stone was cut so that it would protrude a bit from the wall. It would then be chiseled away for a flush fit.

Labor-saving devices such as electric saws, air-powered chisels, and lathes negated a need for a blacksmith to sharpen and replace tools. Modern chisels, with carbide tips, and air-powered tools did not wear out quickly. But every effort was taken to reflect the original

Modern machinery helped workers achieve the look of the original stone yet saved them considerable time.
Author

flavor of the stone. If the stone had a furrowed surface, the replacement stone would, too. Combining modern equipment and traditional tools, a restored and authentic appearance resulted.

Modern equipment was especially helpful in moving heavy replacement stone. One piece weighed 3,000 pounds, requiring a crane. However, most stones were lifted by traditional pulleys upon scaffolding. Modern equipment was also helpful with the replacement of the balusters, the vase-shaped supports for the White House balustrade, or top railing. Mechanical lathes accurately shaped the stone and saved the masons considerable time.

Presidents Reagan and Bush and their families lived and worked amidst the workmen, scaffolding, stone dust, and noise. When asked about the inconvenience, President George H. W. Bush replied that he loved the house so much that he did not mind the inconvenience.[22] President Ronald Reagan wrote:

> Living in the White House those eight years was a gift from the American people that Nancy and I will treasure forever. We both had always shared the reverence most Americans have for that historic building and living there was a joy and the privilege of a lifetime.
>
> Restoration is an important and inevitable process. We never really considered it an inconvenience. Actually, it made us feel good that people were working to preserve the beauty and dignity of "America's home."[23]

RAMCO president Thomas Rudder recommended the exterior facade of the White House receive a washdown every two to three years with a "non corrosive detergent with good rinseabilty." This would remove atmospheric pollutants from the coating and remove any loose or flaking paint. He suggested paint should be added only when the Aquia stone shows through. In 1991 the paint originally applied in 1980 was stripped for repair work. After a decade, the newly stripped paint came off relatively easily. Mr. Rudder wrote, "…the Aquia Creek sandstone was in fine condition, with no change from the initial stripping."[24] Patrick Plunkett oversaw the White House's stone restoration project until its 1996 completion. Today washing and painting proceed when necessary, with full painting occurring about every ten years.

23

Removing and Replacing the Capitol's Stone

Even the stately Aquia stone edifices were not free from Washington's "media circus" and political atmosphere. In 1964, columnist Drew Pearson caused a stir by writing incorrect information about Government Island. He stated that George Washington once owned the quarrying site. He also stated that J. George Stewart, Architect of the Capitol, believed that the Capitol's west wall was in danger of falling and must be reconstructed and reinforced. A portion of the article read:

> Now, the Capitol's west wall (in comparison to the east) is in even more serious need of repair. It is built of sandstone put together with mortar made of burnt oyster shell and placed on almost no foundation…George Washington was a great economizer and a shrewd businessman. He sold the Capitol architect sandstone from Aquia, the island which he owned in the Potomac. It was hauled up on barges from below Mount Vernon and spliced together with cement which cannot be duplicated today.[1]

David Brinkley, based on Pearson's article, disseminated the misinformation on NBC's *Huntley-Brinkley News*. Meanwhile, Mount Vernon's research and reference librarian exposed the truth that George Washington never owned the island and in fact bought stone from the island himself.

Washington, D.C., schoolteacher George Hodgkins wrote to Pearson and Brinkley about the island's true ownership. Letters of appreciation for calling the error to their attention can be found in the Architect of the Capitol's files.

Birthstone of the White House and Capitol

Virginia Senator A. Willis Robertson, reacting to Pearson's report, placed the Richmond's *News-Leader* editorial in the Congressional Record. Part of the article stated:

> Now it was all very peculiar why Pearson should run a smear on George Washington when there was no immediate prospect of Washington's running in an election. But then the political angle cleared up, J. George Stewart is gunning a campaign to enlarge the west front of the Capitol, at the cost of untold millions of dollars. He wants to replace the original Aquia stone with white marble to match the architectural disaster of the east front, which he mangled a few years back. Pearson for some reason has been larding up the campaign with horrendous stories about the deterioration of Aquia stone. The slam at Washington apparently was false moral justification for a project that is architecturally unjustified....[2]

In the 1960s the west facade was shored up with large timber trusses, preventing deteriorating stone from falling. During the 1960s and 1970s, there were periodic discussions as to whether the West Front of the Capitol should be replaced and extended like the East Front or just restored. Due to the outcry of architects, politicians, and preservationists, both projects were abandoned. Then, on April 27, 1983, one hundred square feet of Aquia stone fell from the colonnade, landing in the courtyard. Fortunately, no one was injured. Approximately two months later, the House and Senate agreed to appropriate $49 million for West Front restoration. Rather than rebuild the entire facade, a program would be developed to strengthen, renovate, and preserve the original front. President Reagan signed the legislation on July 30, 1983.[3] Twenty years had passed since the first suggestion of West Front restoration, but it took only three months after the collapse of the stone for something to be done.

Paint was removed using much the same process as the White House. A coat of stripper was applied and remained on the surface of the stone for twenty-four hours. Then another layer of stripper was applied. This, too, remained on for an additional twenty-four hours. Both coats were removed with water. Then an acid solution was applied and stayed on the stone for three to five minutes to neutralize the stone. This, too, was washed off with water. The removal of the paint was completed in February of 1984.[4] The White House uncovered forty-two layers; the Capitol found only thirty-five coats.

PAINT AND THE CAPITOL

The President's Mansion had been painted in 1798. However, the Capitol was painted in 1819, nineteen years after the first section was occupied. Bill Allen, Architectural Historian for the Capitol, developed a theory that the Capitol was never intended to be painted. After paint was removed from the Capitol and the White House in the 1980s, it was discovered that freestone used on the North and South buildings of the "Old Capitol" was unblemished and almost pure white. However, the stone used at the White

House showed iron [ferrous oxide] veins or reddish streaks. He believes that the Capitol was envisioned as a shining example of purity and strength and, therefore, constructed of the best Aqua stone. The only reason it was painted was to cover the damage from the 1814 British burning.[5]

However, Rex Scouten, former White House Curator, believes the White House possesses the better quality stone. Throughout the years, it withstood damage better, perhaps because of the protection given by layers of paint. Thomas Rudder, whose company removed the White House paint, concluded, "Overall, the sandstone was in remarkably good condition with no significant deterioration. Apparently there are some advantages to having 36 to 50 coats of paint on stone."[6]

As at the White House, Capitol paint removal helped assess damage to the stone. Uncovered damage was attributed to:

- Defects in the original foundation
- Deterioration of the Aqua stone
- Fires of 1814 and 1851
- Gas explosion of 1898
- Fourth-floor addition
- Channeling of walls to install interior utilities

Top: This picture, taken in 1984, shows the massive undertaking of reconstructing the West Front, or Mall side, of the Capitol.
Architect of the Capitol

Bottom: The Capitol's West Front was restored from 1983 until 1987. Damaged Aqua stone was replaced with Indiana limestone. Sixty percent of the Aqua stone remains today.
Architect of the Capitol

Damaged stonework that could be repaired was. Forty percent of the Aqua stone was replaced with Indiana limestone. A stone strengthener was sprayed on the remaining Aqua stone. The stone then received a breathable masonry coating that repels water. Damaged carving was replaced, using casts or models of the original. Over one thousand stainless steel tie rods added strength to the Capitol's masonry. After renovation completion in 1987, the West Front was painted to match the marble wings.[7] The extensive project of paint removal, stone removal, stone restoration, stone installation, and repainting was finished in four years.

Epilogue

Margaret Bayard Smith arrived in Washington in 1800 as the blushing bride of Samuel Harrison Smith who had established the first national newspaper printed in America, the *National Intelligencer*. Her arrival in Washington coincided with that of President John Adams and members of Congress as they sat up residence in the new Federal City. During her forty-four years in the district she recorded her observations about Washington life and society. In 1841 she wrote the following words about the White House walls after witnessing four decades of change:

> "Walls," it is proverbially said, "have ears," had they likewise tongues what important, interesting and amusing facts could the walls of the President's House reveal. What a variety of characters, of events, of scenes, recurs to the mind of one who has watched the mutations which have taken place in this dwelling of our chief magistrates! Each successive administration seems like a complete and separate drama performed by new sets of performers. How changed in every respect, both externally and internally is this National Theatre."[1]

Oh, what stories the walls could tell today, after 200 years! The Aquia stone walls are the only architectural components that remain standing since the White House's initial construction during George Washington's administration. The interior was destroyed during the War of 1812 and reconstructed, only to be reconstructed again during Truman's administration.

The exterior walls of the old Capitol have undergone changes. Marble now duplicates the Aquia stone front of the east side, and only 60 percent remains of the west. What if the Aquia freestone interior walls of the Capitol could speak? For forty years the Senate met within the walls of the Old Senate Chamber and heard the great national debates dealing with issues such as slavery. The Aquia walls of the Old House of Representatives, now Statuary Hall, heard the Marquis de Lafayette address Congress and saw Presidents James Madison, James Monroe, John Quincy Adams, Andrew Jackson, and Millard Filmore inaugurated. For forty-one years the walls in the Old Supreme Court Chamber heard landmark decisions becoming law. The warm buff-colored walls of the Rotunda have observed the laying in state of presidents and eminent citizens.

Truly, the Aquia freestone walls of America's two most important structures have heard and witnessed historical events and changes. Today they shine as symbols of freedom for the world.

Rock of Ages
Recent Recognition and Preservation Efforts

Paper Leads to Preservation, 1979

In 1979, the author wrote a paper about Government Island for independent study credit. After discovering that the island was divided into lots for development, the paper concluded by emphasizing that the island should be saved for its historic significance. After reading the paper, Francis Eck, the lawyer for both the Aquia Harbour Property Owners and for Westinghouse (the owner of development), agreed that it should be saved. He purchased the island for $10 and services rendered.

Government Island up for Sale, 1992

In 1992, it was discovered that Mr. Eck wished to sell the property. He was then working in Richmond and wanted to sell the property for $410,000. After some negotiations, he lowered the price to $400,000, then to $240,000. He said he would wait until the money was raised. Various organizations (such as Aquia Harbour Property Owners, National Park Service, Conservation Fund, and even the White House) were canvassed, but none had the available funds.

Four years later, in October of 1996, a newspaper article appeared with the bold headline, "This Island for Sale." The owner had placed the island for sale by a local Realtor for $400,000. Even with front page coverage, the island failed to find a buyer.

Island Is Purchased, 1998

C. M. Williams, Stafford County Administrator, and several members of the Board of Supervisors traveled to Richmond to negotiate purchase of the island for $200,000. On August 19, 1998, the Stafford County Board of Supervisors purchased the island to preserve it for future generations.

Government Island Committee and Report, 1999–2002

On July 13, 1999, Stafford's Board of Supervisors established a committee to determine the best utilization of the newly acquired island. **Rex Scouten**, who had worked for ten presidents as Secret Service agent, chief usher, and White House curator, was chosen as chairman. He chose the following people to be on the committee:

Charles Atherton, secretary of the U.S. Commission of Fine Arts since 1965 and Fellow of the American Institute of Architects

Appendix I

Jane Conner, member of Stafford Historical Commission, former president of the Stafford Historical Society, and Government Island researcher

William L. Ensign, Assistant Architect of the Capitol for fifteen years and Acting Architect of the Capitol

Dr. Robert Kapsch, Senior Scholar in Historic Architecture and Engineering of the National Park Service and former Chief of HABS/HAER

Robert Stanton, Director of the National Park Service, represented by Marcia Keener, program analyst, Office of Policy, National Park Service

William Seale, author of many books about the White House and also consultant on historic houses

H. Alexander Wise, director of the Virginia Department of Historic Resources, was represented by Calder Loth, senior Architectural Historian

C. M. Williams, Jr., Stafford County Executive and assisted by Wendy Mallow, administrative assistant

On January 30, 2002, the Committee presented a report concluding that the island could best be preserved by continuing Stafford County's ownership and by establishing a low-impact park easily accessible either from Washington, D.C., or Richmond, Virginia. The committee hoped the county would share this unique historical resource with residents and tourists. Other recommendations in the report included identifying archeological opportunities and establishing interpretive signs, trails, etc. The committee also stressed the need for protective preservation designation by state and federal authorities.

TV Spot: African American Labor Used in Capitol
March 2000

Ed Hotaling, noted author and news producer at NBC's Washington, D.C., affiliate, Channel 4, ran a piece disclosing the fact that slaves had been instrumental in creating the most important symbols of freedom in the United States. While researching his spot for the 200th anniversary of the Capitol, he uncovered pay slips showing that out of 650 workers on the building of the Capitol, 50 were free blacks, and up to 400 were slaves. He told the viewing audience that slaves did not receive their wages, only their owners.

Congress Approves Legislation to Honor Slave Laborers
October 2000

Rep. J. C. Watts (R-Okla.) and Rep. John Lewis (D-Ga.) introduced legislation which would honor the African Americans, slave and free, who helped build the U.S. Capitol. Later, on the Senate side, they were joined by Senators Spencer Abraham (R-Mich.) and Blanche Lincoln (D-Ark.). In October, the legislation was approved establishing a task force that would study the contributions and history of slave laborers used in the Capitol's construction. An appropriate memorial was also recommended for the grounds of the Capitol. As of May 2005, the task force was not yet fully constituted.

White House Historical Association's Christmas Ornament 2000

In commemoration with the 200th anniversary of the completion of the White House, the annual Christmas ornament was made with Aquia stone. Due to the generosity of the Stafford County Administration, the association was allowed to collect buckets of loose stone. The stone was ground and mixed with resin. It was then poured into molds of the White House and when cooled, later affixed to brass ornaments. Millions of people across the nation are able to have a little piece of Aquia stone of their own.

National Recognition for Government Island, 2001–2002

Virginia 1st District Congresswoman Jo Ann Davis introduced a Congressional Resolution in October of 2001 that sought national recognition of Government Island.

Suzanne Carr, Fredericksburg Free Lance-Star

Appendix I

The resolution highlighted the significance of the island and stressed its role in providing the building material for both the White House and Capitol.

Four months later, on February 7, 2002, Congresswoman Davis appeared before the House Subcommittee on National Parks, Recreation, and Public Lands to introduce Rex Scouten, former White House curator, and the author. Speaking on Capitol Hill, we stressed the national importance of the island and its stone.

On April 30, 2002, Congresswoman Davis appeared before the House of Representatives for final voting on the resolution. The house unanimously approved. Government Island received federal recognition as a "historical site."

Stone House Preserved, 2002

In 1806 Benjamin Henry Latrobe visited the Robertson Quarry in Stafford. (See Chapter 12.) He spent a night in Robertson's home. In the 1990s, Austin Ridge developers purchased many acres including the foundation of this house and its chimney. Isolated in the woods, next to the foundation, stood an Aquia stone shell of a much larger house known as the Towson House, ca. 1820. Thomas Towson was a stone mason from Baltimore who helped create Baltimore's Washington Monument (coincidentally along with Robert Steuart's son, William, in 1815). Close to 700 homes have been built around the historic house sites and on the quarry site. Fortunately, Rick Wolff, executive vice president of the developing company, decided to save Robertson's foundation and chimney along with the two-and-a-half-story stone shell of the Towson House. He hired Shelton Alley, who had worked on the restoration of the Capitol's West Front in the 1980s. The site, currently known as the Robertson-Towson House, was reinforced. A wrought iron fence was placed around the property so residents could view the historic area without fear of injury and read signage about its history. This preservation effort cost the development company $100,000, but residents can now realize the important role Stafford played in the building of the nation's capital city.

Now hundreds of houses in Austin Ridge subdivision surround the preserved Robertson-Towson home sites.
Jack Boucher, Historic American Buildings Survey

Appendix I

Artist Visits Island, 2003

Noted British environmental artist and sculptor Andy Goldsworthy visited Government Island at the request of the National Gallery of Art in D.C. Goldsworthy, who regards all of his creations as temporary, photographed nature upon the island. His work, created on the isle, is included in his book, *Passage*.

Landmark Status, 2003

On March 19, 2003, Government Island was named to the Virginia Landmarks Register. Aaron Shriber, then Stafford's historic preservation planner, completed paperwork on the island. After the island was placed on the Commonwealth's Register, it was nominated to be on the National Register of Historic Places.

Remembering 9/11, New York 9/11/02

To honor the memory of September 11 victims and to commemorate the heroism of countless others, 50 senators and 250 members of the House of Representatives traveled to New York a year after the tragedy for a ceremony at Federal Hall.

To commemorate the ceremony, a plaque was presented to Federal Hall. It was affixed to Aquia stone. The stone was some of what had been removed during the Capitol's eastern extension. Representative Vito Fossella from New York told of the Aquia stone's significance.

(Few watching the proceedings on television realized the importance for Stafford County. First of all, the stone was from Stafford. Second, George Washington, who grew up in Stafford, was sworn in as President at Federal Hall. Third, George Mason, who also grew up in Stafford, wrote the Virginia Declaration of Rights, the basis for our nation's Bill of Rights. The Bill of Rights was written at Federal Hall.)

Nixon Library, 2003

Stafford County contributed a rectangular block of Aquia stone to the Richard Nixon Library and Birthplace in Yorba Linda, California. It was used as a cornerstone for its library expansion. Part of the expansion included re-creation of the White House's East Room. The $12 million project was made possible by Katherine B. Loker, whose father was the founder of the Star Kist Foods Company. Today the wall outside the East Room displays the block graced with a plaque saying that it came from Government Island in Stafford County, Virginia.

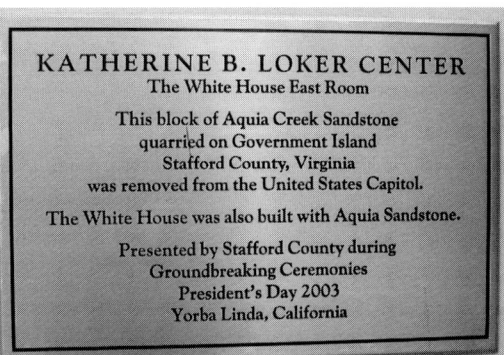

Appendix I

Kenmore, 2004

George Washington's Fredericksburg Foundation

Kenmore, the Fredericksburg home of George Washington's sister and brother-in-law, Betty and Fielding Lewis, underwent a restoration project in the early 2000s. During that time it was discovered that the original eighteenth-century sandstone steps on the west side of the house needed to be replaced. Since Government Island could not be used to obtain the stone, a newspaper article was written mentioning the need for freestone. Fortunately, in 2004, Stafford County was widening a road and uncovered a bed of the stone. The county provided Kenmore with twelve huge boulders ranging in weight from about five to thirty-nine tons. The largest boulder was larger than a minivan. Some boulders were sent to North Carolina, where they were shaped into slabs to rebuild the steps.

History of Slave Labor, 2005

In early 2005, a Capitol Appropriations Committee directed the Architect of the Capitol to study the contribution of slave labor at the Capitol. In June, William Allen, Architectural Historian for the Architect of the Capitol's office, created a document titled *History of Slave Labor in the Construction of the U.S. Capitol*, in which he mentions that the creation of the buildings was the first time slaves were used in the capital city.

Opening Government Island, 2005–2006

Stafford County's Department of Economic Development awarded a contract to a firm for engineering and designing low-impact trails to Government Island. Construction should begin the spring of 2006 with a target opening of the island to public access in late 2006.

Appendix II

OTHER STRUCTURES OF AQUIA STONE

In addition to the White House and United States Capitol, other structures used freestone in their construction. Most do not have full documentation stating they were built with stone from Government Island or other quarries along Aquia Creek. A logical assumption, however, is that structures built prior to 1791 were constructed from island stone, for the Brents had been supplying the area with stone for decades. After that date, when L'Enfant purchased the island, most stone went for construction of the Federal City's two most important buildings. Other quarries along Aquia Creek opened up during this period. A few remained in operation into the early 1900s.

The following buildings or structures are listed in chronological order of their construction:

THE NELSON HOUSE
Yorktown, Virginia
Ca. 1730

On a bluff in Yorktown, Virginia, this stately mansion was constructed of English-made bricks, previously used as ships' ballast. Aquia stone was used for decorative trim. During the Revolutionary War, Thomas Nelson, Jr., lived in the house that was built by his grandfather. Nelson, a signer of the Declaration of Independence, war-time governor of

Author

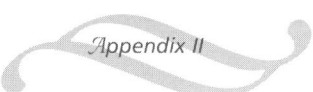

Virginia, and General of the Virginia Militia, had troops fire on his own home when he realized that Cornwallis had taken over the structure. Cannonballs can be found embedded in the brick walls.

Author

AQUIA CHURCH
Stafford, Virginia
1751

Aquia Church, an excellent example of combining freestone and brick in colonial Virginia, was built in cruciform design; its eight corners were trimmed with freestone quoins. Keystones of Aquia stone were inserted in the round brick arches of the windows. Doorways were decorated with sandstone instead of the rubbed brick popular in other Tidewater churches.[1] Since it is located close to Aquia Creek, the Aquia stone was probably quarried from Brent's Island. It was built in 1751, forty years prior to the island's purchase by L'Enfant.

Appendix II

Author

JOHN CARLYLE HOUSE
Alexandria, Virginia
1753

This Georgian Palladian-style manor house, overlooking the Potomac River, was completed in 1753 by John Carlyle, a wealthy Scottish merchant. Originally constructed entirely out of Aquia stone, it was the only solid stone house in Alexandria at the time. In 1755, British General Braddock occupied Carlyle's house as headquarters to plan the early French and Indian War campaigns. Today the Aquia stone on the exteriors is not visible, for the back and sides were covered in stucco and the front façade was covered with a thin veneer of Indiana limestone.

GUNSTON HALL
Fairfax County, Virginia
1755–1758

Gunston Hall was constructed from 1755 to 1758. It was the home of George Mason, author of the Virginia Declaration of Rights, the forerunner of the Bill of Rights. In the Georgian home of brick, laid in Flemish bond, freestone was used in quoins, chimney

Appendix II

Gunston Hall.
Jack E. Boucher, Historic American Buildings Survey/Library of Congress

Mount Airy.
Jack E. Boucher, Historic American Buildings Survey/Library of Congress

caps, the water table, and as keys to the flat arches. No documentation exists that the stone came from Brent's Island, but as Mason was raised in Stafford County, he knew of its existence. (Mason's second wife was Sarah Brent, daughter of George Brent of Woodstock.) Mason also suggested his son look at stone from the Aquia quarries for chimney pieces, or mantels.[3]

MOUNT AIRY
Richmond County, Virginia
1757

Unlike most colonial Virginia plantation houses, Mount Airy was constructed of gray sandstone indigenous to the area. White sandstone from Aquia Creek was used for contrasting trim. It can be seen in string courses, or horizontal bands, quoins, window enframements, rusticated center pavilions, and pedestals at the top of terrace steps that support a pair of stone urns. This stately Georgian home, completed in 1757, was designed by its owner, John Tayloe III, and has remained in the Tayloe family for over 200 years. Mrs. Henry Gwynne Tayloe, Jr., related that Aquia stone traveled by boat from Aquia Creek via the Potomac to the Rappahannock to a wharf close to the house.[4] Fiske Kimball, noted architectural author, wrote that Mount Airy was "perhaps the most ambitious home in the colony."[5] (See "The Octagon," on page 183, for another Tayloe home.)

CHRIST CHURCH
Alexandria, Virginia
1767–1773

Original pews in which George Washington and General Robert E. Lee worshiped are marked within this Georgian building. Built from plans by noted colonial architect James Wren, the church was constructed of red brick and accented with Aquia stone quoins and keystones. Unlike the freestone in Aquia Church and Pohick Church, its stone is painted white. The church is a major tourist attraction in downtown Alexandria.

Author

Appendix II

Williams Ordinary.
Author

Pohick Church.
Author

Appendix II

WILLIAMS ORDINARY
Dumfries, Virginia
1760s

This eighteenth-century building, known in colonial times as Williams Ordinary, went by later names of Love's Tavern, Stage Coach Inn, and Old Hotel. The structure is architecturally significant, as it appears to be the only remaining Virginia building with a front wall laid in all-header bond. Aquia trim can be seen on the quoins and doorway. It is suggested that perhaps the structure was built around 1765 by James Wren, based on architectural similarities between Christ Church in Alexandria and this structure.

POHICK CHURCH
Fort Belvoir, Virginia
1769–1774

Aquia stone can be seen on the carved pedimented doorways and quoins of this church erected from 1769 to 1774. William Copein was the church's mason and is believed to be responsible for the carved pediment doorways. Both George Washington and George Mason were vestry members and played roles in its construction. Its design is attributed to James Wren, as the Fairfax Parish paid him for drawings of Pohick Church, Christ Church in Alexandria, and Falls Church in Falls Church, Virginia. There are definite similarities in these structures.[6]

KENMORE
Fredericksburg, Virginia
1770s

This elegant Georgian mansion was home to Betty Lewis, George Washington's sister and the wife of wealthy plantation owner Colonel Fielding Lewis. The red brick house was built between a pair of detached wings around 1775. The portico, facing the Rappahannock River, was made with Aquia Stone. The classic simplicity of the exterior is in contrast with the ornate plaster ceilings and chimney pieces of the interior. Lewis aided his brother-in-law and the American cause by investing much of his personal fortune in arms and ammunition for the Revolutionary War. Never repaid, his estate had to be sold.[2]

The Kenmore.
Jack E. Boucher, Historic American Buildings Survey/Library of Congress

CAPE HENRY LIGHTHOUSE

U.S. Route 60, Virginia Beach, Virginia
1790–1792

Cape Henry Lighthouse was the first lighthouse commissioned by the First Congress of the United States of America. American-born architect John McComb, Jr., was offered the contract. Constructed of hammered-dressed Rappahannock sandstone ashlar, the tower is seventy-two feet high with a spread foundation of twenty feet deep. Aquia stone was used on foundations, windows, and door surrounds. The governor of Virginia on July 18, 1791, discussed McComb's difficulty in completing the lighthouse:

Cape Henry Lighthouse
APVA

> Mr. McComb is hard at work raising the stone, according to contract. But he has great difficulty because of the sand, which rolls back into the excavations almost as fast as he removes it. This has to be done by means of wheel-barrows. The sand frequently has to be carried fifty yards in order to facilitate the work, &c.[7]

This octagon-shaped lighthouse, completed in October of 1792, took nineteen months to construct.[8]

Appendix II

THE OCTAGON
18th & New York Avenue, NW
Washington, D.C.
1798

The Octagon.
Jack E. Boucher, Historic American Buildings Survey/Library of Congress

This unique structure was designed by Dr. William Thornton, first architect of the United States Capitol, in 1798. Built as a winter season townhouse for Colonel John Tayloe III of Mount Airy, Virginia, it predates the neighboring White House. In 1814, after the British burned the President's Mansion, President James Madison and wife Dolley lived there. The Treaty of Ghent, ending the War of 1812, was ratified within this house. Like the Tayloe home, Mount Airy, this building has Aquia stone trim that can be found in horizontal bands and exterior plaques. Today it is called the Octagon Museum and is considered to be the oldest museum in the United States devoted to architecture and design.

WOODLAWN PLANTATION
U.S. Highway 1
Mount Vernon, Virginia
1805

As a wedding gift, George Washington presented granddaughter Nelly Custis and nephew Lawrence Lewis with 2,000 acres. Included in the tract was Gray's Hill, an area

Woodlawn Plantation.
Jack E. Boucher, Historic American Buildings Survey/Library of Congress

Appendix II

the president marked off as "a most beautiful Site for a Gentleman's Seat." The couple stayed at Mount Vernon until after the death of both General and Mrs. Washington. While living there, their new home, named Woodlawn, was built upon Washington's suggested site. In 1802, the Lewises moved into the house; however, it was not completed until 1805. Woodlawn was designed by Dr. William Thornton. Like the Octagon, also designed by him, Woodlawn has Aquia stone trim. It can be seen in sills and lintels of the windows and doors, and basement coping. Lawrence Lewis was familiar with Aquia stone, for it was also used in his childhood home, Kenmore.

Courtesy Congressional Cemetery

CONGRESSIONAL CEMETERY
Washington, D.C.
1812

In 1812 Congress bought one hundred plots from the Washington Parish Burial Grounds to inter members who died in office and whose remains could not be moved to their native states. Fourteen senators and over sixty representatives were buried in the cemetery.[9] The Congressional Cemetery, as it is now called, has quite a few freestone markers. Aquia stone was shipped to Baltimore and Philadelphia for tombstones, so perhaps stone was shipped for these memorials, or unused stone from the federal buildings was utilized.

Monumental Church.
Jack E. Boucher, Historic American Buildings Survey/Library of Congress

MONUMENTAL CHURCH
Richmond, Virginia
1814

This octagonal-shaped church was built as a memorial for seventy-two people who perished in an 1811 Richmond Theatre fire. Chief Justice John Marshall headed fundraising efforts to memorialize the victims. Robert Mills, designer of the Washington Monument, designed the church. Built upon the site of the theater, the walls were made of Aquia stone.

C & O CANAL
Georgetown
1828

Construction of the Chesapeake and Ohio canal was begun in 1828. Since much stone was needed, existing and new quarries were used. Most of the seventy-four locks were of granite or red Seneca Creek Sandstone. However, two locks used a great deal of dressed Aquia stone, and two other granite locks had Aquia Creek freestone for coping. Aquia stone was used in the construction of five Georgetown bridges.[10] (Of the five bridges, the Wisconsin Avenue Bridge, which is faced with freestone, is the only one standing today.) The canal, completed in 1850, spanned approximately 185 miles.

Library of Congress

FORTRESS MONROE
Old Point Comfort
Hampton, Virginia
1819–1835

This moated fortification was built between 1819 and 1835. The fort is constructed mostly of granite, but the Main Sally Port is composed of sandstone. Some of the brick

Fortress Monroe.
Jack E. Boucher, Historic American Buildings Survey/Library of Congress

Appendix II

residences located within the fort and just outside are trimmed with freestone. Records from the National Archives list the ships that carried the freestone but not the quarries from which the stone was extracted. Oral tradition credits the Aquia Creek region.[11]

WHITE HOUSE STABLES
1833–1834

President Andrew Jackson had many fine race horses, which were kept in the west wing and wooden shanties on White House grounds. It was said that odors from the horses flowed into the State Dining Room. In order to prevent this and to better house his horses, Jackson had a new stable built in 1833–34. The stable was made of bricks covered with stucco. Aquia stone trim graced the windows. It was located southeast of the White House, where the statue of General William T. Sherman now stands.[12]

OLD PATENT OFFICE
Washington, D.C.
1836–1867

The "Old Patent Office" was started in 1836 and completed in 1867. Only the south wing is of sandstone. Other wings are of white marble. The south wing façade is patterned after the Parthenon. In 1964–67 this Greek Revival building became home of the National Portrait Gallery and the National Museum of American Art, but it had many prior occupants. Walt Whitman worked there when it housed the Department of the Interior. President Abraham Lincoln was there for his second inaugural. Clara Barton nursed wounded soldiers in the building after the outbreak of the Civil War. Before the completion of the National Archives, it housed the Declaration of Independence and other historic documents.[13]

Old Patent Office.
Ronald Comedy, Historic American Buildings Survey/Library of Congress

UNITED STATES DEPARTMENT OF TREASURY
Washington, D.C.
1836–1869

Two previous Treasury buildings were destroyed by fire. This structure was begun in 1836 and required thirty-three years to complete. During construction in 1845, William

Appendix II

U.S. Department of Treasury.
*Jack E. Boucher, Historic American Buildings Survey/
Library of Congress*

Force wrote in the Wm. Q. Force Guide, "a noble structure - pity it were not built of something more durable than sandstone...."[14] Today it is completely made of granite. The last sandstone façade, on the east side, was replaced with granite in 1908.

TWENTIETH-CENTURY USE OF AQUIA STONE

Files of the Architect of the Capitol contain a list of structures that used Aquia stone. The list dates to May of 1930. Written in pen is a note, "Within the last five years." The only stone quarry in operation on Aquia Creek during 1925 through 1930 was the George Washington Stone Corporation. Many of the buildings on the list no longer exist; others have had name changes. The list suggests that the stone had been delivered as far south as North Carolina, as far north as Connecticut, and as far west as Ohio.

Employees of Aquia Creek's George Washington Stone Corporation gather in front of a channeling machine in 1929.
Courtesy Wilbur Segar

Appendix II

Harkness Hall, Yale University, New Haven, Connecticut
Gallery of Fine Arts, New Haven, Connecticut
Knights of Columbus Home, Baltimore, Maryland
Provident Savings Bank Branch, Baltimore, Maryland
S. & N. Katz Building, Baltimore, Maryland
Fort Lincoln Chapel, Washington, D.C.
Chesapeake & Potomac Telephone Exchange, Washington, D.C.
St. Francis de Sales Church, Washington, D.C.
Alta Vista Apartments, Washington, D.C.
Fourth Presbyterian Church, Washington, D.C.
United Publishing Company Building, Washington, D.C.
National Bible Institute, New York, New York
National Bank of North Philadelphia, Pennsylvania
Jewish Community Center, Philadelphia, Pennsylvania
Preis Building, Philadelphia, Pennsylvania
Weinrath Building, Philadelphia, Pennsylvania
U.G.I. Office Building, Philadelphia, Pennsylvania
Bryn Mawr College Dormitories, Bryn Mawr, Pennsylvania
Moorestown Trust Company, Moorestown, New Jersey
Lafayette Hotel, Atlantic City, New Jersey
Planters National Bank, Fredericksburg, Virginia
T. C. Williams Residence, Richmond, Virginia
Telephone Exchange, New York, New York
Richmond Public Library, Richmond, Virginia
J. H. Adams Residence, Sedgefield, North Carolina
Drug Store, Fredericksburg, Virginia
Kishpaugh Store Front, Fredericksburg, Virginia
Jefferson Hospital, Philadelphia, Pennsylvania
Superintendent's Residence, Eastern State Hospital, Williamsburg, Virginia
Telephone Exchange, Hopewell, Virginia
Potomac Electric Power Company Sub-Station, Washington, D.C.
Store fronts for Beveridge Estate, Washington, D.C.
McCormick Residence, Washington, D.C.
Laughlan Residence, Washington, D.C.

Appendix II

Maury School, Alexandria, Virginia

Snyder, Kane, Booth Corporation Building, Alexandria, Virginia

Cemetery Plot, Troy, New York

John Quincy Adams School, Washington, D.C.

Wineman Building, Washington, D.C.

Augustus Residence, Cleveland, Ohio

Kenmore, Fredericksburg, Virginia

Bishop's Gardens, Washington Cathedral, Washington, D.C.

Cassidy Office Building, Washington, D.C.

Pyne Residence, East Orange, N.J.

Auto Show Room, Washington, D.C.

Aspinwall Residence, Washington, D.C.

St. Gertrudes School, Washington, D.C.

Dr. Latimer Residence, Washington, D.C.

Gunston Hall Gardens, Gunston Hall, Virginia

Donovan Residence, Washington, D.C.

Washington Gas Light Company, Washington, D.C.

Wire Residence, Washington D.C.

Dr. Shields Memorial, Washington D.C.

Georger Residence, Warrenton, Virginia

Diener Residence, Alexandria, Virginia

Lambert Residence, Alexandria, Virginia

Olive Dann Residence, New Haven, Connecticut

Flag Pole Base, Episcopal High School, Alexandria, Virginia

Lee Residence, Washington, D.C.

Frederick Scott Country Place, Afton, Virginia

Harper Residence, The Plains, Virginia

Weigand Residence, Washington, D.C.

Davis Residence, Beane, Maryland

Woodstock College, Woodstock, Maryland

Finlinson Residence, Richmond, Virginia

Remey Residence, Washington, D.C.

Calvary Baptist Church, Washington, D.C.

Appendix II

FREDERICKSBURG AREA MUSEUM AND CULTURAL CENTER
Fredericksburg, Virginia
1927

This baroque-style building is found on the above list. Built as Planters National Bank, it later went by the names of Farmers and Merchants and First Virginia. In 2004 BB&T, which acquired the property, worked with the neighboring museum for acquisition.

Author

PILGRIM STEPS
WASHINGTON NATIONAL CATHEDRAL
Washington, D.C.
1920s

In 1893 the Episcopal Church was chartered by Congress to build "a house of prayer for all people." Started in 1907 and completed in 1990, this Gothic place of worship is the sixth-largest cathedral in the world. The church walls are made of Indiana limestone, but

Courtesy National Cathedral

the fifty-one broad steps leading to the south transept entrance, known as Pilgrim Steps, were made of Aquia stone in the 1920s. The walls surrounding the Bishop's Garden, modeled after a medieval walled garden, were also made of freestone. For decades, docents told visitors that the stone for the steps and walls was from George Washington's own quarry. In 1994 the docents discovered that the stone was acquired from the George Washington Stone Corporation, not the President's quarry.[15]

JOSLYN ART MUSEUM
Omaha, Nebraska
1931

Seven years after completion, the Joslyn Art Museum was listed as one of the one hundred finest buildings in the United States. Most of the three-level interior was constructed from thirty-eight different marbles from around the world. The walls of the Fountain Court were made of Aquia Stone. Construction of this $3 million museum took place from 1929 to 1931. As the George Washington Stone Corporation was the only Aquia stone quarrier in existence during this time, the stone must have come from its Aquia quarry. (This Art Deco museum was not on the above list, for it was completed after the list was created. It does indicate that the George Washington Stone Corporation expanded its shipping region as far west as Nebraska.)

Courtesy Joslyn Art Museum

otes

The following abbreviations will be found in the Notes:

GPO: Government Printing Office
LOC: Library of Congress
NA: National Archives
P: Press
U: University

Prologue

[1] Abigail Adams, *New Letters of Abigail Adams, 1788–1801*, ed. Stewart Mitchell (Boston: Houghton Mifflin Co., 1947), 259.

[2] Allan Nevins, *We, the People, 13th ed.* (Washington: U.S. Capitol Historical Society with National Geographical Society, 1985), 6.

1. What Is Aquia Stone?

[1] John Smith, *The Complete Works of Captain John Smith, 1580–1631*, vol. 1, ed. Philip L. Barbour (Chapel Hill: U of North Carolina P, 1986), 146–48.

[2] F. J. Pettijohn, Paul Edwin Potter, and Raymond Siever, *Sand and Sandstone* (New York: Springer-Verlag, 1972), 158.

[3] Marvin F. Studebaker, "Freestone from Aquia," *Virginia Cavalcade*, 9, no. 1 Summer 1959: 36.

[4] Land patent, November 28, 1678, Washington: NA.

[5] Deed, 30 January 1694, Washington: NA.

[6] *The Register of Overwharton Parish, Stafford Virginia 1723–1758*, Compiled by George Harrison Sanford King (Easley, S.C.: Southern Historical P, 1961), 169–78.

[7] Bruce E. Steiner, "The Catholic Brents of Colonial Virginia," *Virginia Magazine of History and Biography*, LXX, October 1962: 399–403.

[8] King 130.

[9] Henry Irving Brock, *Colonial Churches in Virginia* (Port Washington: Cale P, 1930), 4.

[10] George Mason, *The Papers of George Mason, 1725–1792*, vol. 1, ed. Robert A. Rutland (Chapel Hill: U of North Carolina P, 1970), 1271–72.

[11] J. D. Morgan, "Major Robert Brent, First Mayor of Washington City," *Records of the Columbia Historical Society*, vol. 2 (Washington: Columbia Historical Society, June 1897), 238.

2. Aquia Stone Boundary Markers for the New Capital

[1] Kenneth R. Bowling, "The Other G. W.," *Washington History*, vol. 3, no. 2, Fall/Winter 1991–1992: 5.

[2] John Ball Osborne, "The First President's Interest in Washington as Told by Himself," *Records of the Columbia Historical Society*, vol. 4 (Washington: Columbia Historical Society, 1900), 178–79.

[3] Silvio A. Bedini, *The Life of Benjamin Banneker* (Rancho Cordova, CA: Landmark Enterprises, 1972), 124.

[4] Charles Callahan, *Washington the Man and the Mason* (Washington: Gibson Bros., 1913), 290.

[5] Fred E. Woodward, "A Ramble along the Boundary Stones of the District of Columbia with a Camera," *Records of the Columbia Historical Society*, vol. 10 (Washington: Columbia Historical Society, 1907), 66.

[6] *Boundary Markers of the Nation's Capital—A National Capital Planning Commission Bicentennial Report* (Washington: GPO, Summer 1976), 37.

[7] Bill Gifford, "On the Borderline," *City Paper*, vol. 13, no. 10, 12–18 March 1993: 24.

3. Who Chose the Stone?

[1] James Henry Heron, *A Historical Sketch of Fredericksburg Lodge no. 4* (Fredericksburg: Kishpaugh P, 1932), 15–16.

[2] George Washington, Ledger B, f. 110, Washington, D.C., LOC, 28 April 1774, and 3 May 1774.

[3] George Washington, *The Diaries of George Washington*, ed. Donald Jackson and Dorothy Twohig, vol. 4 (Charlottesville: UP of VA, 1978), 264–65.

[4] "Writings of Washington Relating to National Capital, *Records of the Columbia Historical Society*, vol. 17, (Washington: Columbia Historical Society, 1914), 229–30.

[5] *Records of the District of Columbia Commissioners*, Letters received, Washington to Commissioners, 23 July 1794, NA Microfilm Publication M-371.

[6] Commissioners of the District of Columbia, Reel 1, Shelf 18,008, LOC. June 28, 1796 (Washington declared October 17, 1791).

[7] Kenneth R. Bowling, *Peter Charles L'Enfant...* (Washington: The Friends of the GW Libraries, 2002), 1.

[8] H. Paul Caemmerer, *The Life of Pierre Charles L'Enfant* (Washington: National Republic P, 1950), 127.

Notes

[9] Caemmerer 128.

[10] Caemmerer 146.

[11] Caemmerer 147.

[12] Caemmerer 178.

[13] Deeds, 2 December 1791, and 3 February 1792, Washington: NA.

[14] Land Deed, 8 May 1786, Washington: NA.

[15] *Records of the District of Columbia Commissioners*, Proceedings, November 1791, NA Microfilm Publication M-371.

[16] *Records of the District of Columbia Commissioners*, Proceedings, 2 January 1798, NA Microfilm.

[17] Bob Arnebeck, *Through a Fiery Trial* (Lanham, MD: Madison Books, 1991), 89.

4. L'Enfant and the City's Genesis

[1] H. Paul Caemmerer, *The Life of Pierre Charles L'Enfant* (Washington: National Republic P, 1950), 152.

[2] Elizabeth S. Kite, *L'Enfant and Washington* (Baltimore: Johns Hopkins P, 1929), 18–19.

[3] George Washington, *Writings of Washington*, ed. John C. Fitzpatrick, vol. 32 (Washington, D.C.: U.S. GPO, 1939), 94.

[4] Caemmerer 155.

[5] John Ball Osborne, "The First President's Interest in Washington as Told by Himself," *Records of the Columbia Historical Society*, vol. 4 (Washington: Columbia Historical Society, 1901), 198.

[6] Caemmerer 187.

[7] William Tindall, *Standard History of the City of Washington from a Study of the Original Sources* (Knoxville, Tenn.: H. W. Crew & Co., 1914), 136–38.

[8] Kite 117–20.

[9] Kite 174.

[10] Caemmerer 196–97.

[11] Kite 27.

5. Jefferson, Brick vs. Stone

[1] Thomas Jefferson, *The Writings of Thomas Jefferson*, ed. Andrew A. Lipscomb, vol. 19 (Washington: Thomas Jefferson Memorial Association, 1903), 90.

[2] *Records of the District of Columbia Commissioners*, Letters received, 13 March 1792, Jefferson's draft for advertisement, NA, M-371.

[3] George Washington, *Writings of Washington*, Letter from Washington to Commissioners, ed. John C. Fitzpatrick, vol. 32 (Washington D.C.: U.S. GPO, 1939), 52.

[4] William Seale, *The President's House*, vol. 1 (Washington: White House Historical Society with cooperation of National Geographic Society, 1961), 28.

[5] Seale, *President's House*, vol. 1, 44–46.

[6] Washington, *Writings*, vol. 32, 325.

[7] Constance McLaughlin Green, *Washington Village and Capital 1800–1878* (Princeton: Princeton UP, 1962), 69.

[8] John Stewart, "Early Maps and Surveyors of the City of Washington, D.C.," *Records of the Columbia Historical Society,* vol. 2 (Washington, 1895), 68–69.

6. Cornerstone Ceremonies

[1] Bob Arnebeck, *Through a Fiery Trial; Building Washington 1790–1800* (Lanham: Madison Books, 1991), 129.

[2] *Building Stones of our Nation's Capital*, U.S. Department of the Interior Geological Survey (Washington: U.S. GPO, 1975), 5.

[3] *Charleston City Gazette*, 15 Nov 1792, Charleston Library Society, Charleston, SC.

[4] *Records of the District of Columbia Commissioners*, Letters received, Hoban and Williamson to Commissioners, 22 June 1793, NA, M-371, no 189.

[5] William C. Allen, personal interview, 18 August 1993.

[6] *Records of the District of Columbia Commissioners*, Proceedings, State of the Public Buildings, 2 September 1793, NA, M-371, no. 194.

[7] H. Paul Caemmerer, "The Sesquicentennial of the Laying of the Cornerstone of the United States Capitol by George Washington," *Records of the Columbia Historical Society* vol. 44–45, ed. Newman F. McGirr (Washington: Columbia Historical Society, 1944), 176.

[8] Wilhelmus Bogart Bryan, *A History of the National Capital*, vol. 1 (New York: MacMillan Company, 1914), 213.

[9] Arnebeck 175.

[10] Caemmerer 177.

[11] Conversation with Dustin Smith, curator of the George Washington National Memorial, Alexandria, Virginia. 22 June 05.

Notes

[11] S. Brent Morris, *Cornerstones of Freedom; A Masonic Tradition*, ed. John W. Boettjer (Washington: The Supreme Council, 33°,S. J., 1993), 46, 73–75.

[12] Caemmerer 178.

[13] John W. Reps, *Washington on View*, The Nation's Capital Since 1790 (Chapel Hill, U of North Carolina P, 1991), 44.

7. Many Workers Are Needed

[1] Harley J. McKee, *Introduction to Early American Masonry: Stone, Brick, Mortar and Plaster* (Washington: National Trust for Historic Preservation, 1980), 20.

[2] James Stevens Curl, *The Art and Architecture of Freemasonry. An Introductory Study* (London: B. T. Batsford Ltd., 1991), 21.

[3] James Stevens Curl, address, "The Capitol in Washington, D.C., and Its Freemasonic Connections," U.S. Capitol Historical Society Conference, Washington, 11 March 1993.

[4] Paul F. Norton, *Latrobe, Jefferson and the National Capitol* (New York: Garland Publishing, Inc., 1977), 262.

[5] McKee 20.

[6] William Tindall, *Standard History of the City of Washington from a Study of the Original Sources* (Knoxville, Tenn.: H. W. Crew & Co., 1914), 182–85.

[7] Tindall 180.

[8] *Records of the District of Columbia Commissioners*, Proceedings, 13 April 1792, NA Microfilm Publication M-31.

[9] Wilhelmus Bogart Bryan, *A History of the National Capital 1790–1814*, vol. 1 (New York: MacMillan Co., 1914), 232.

[10] Julian Niemcewicz, *Under Their Vine and Fig Tree; The Travels Through America 1797–1799, 1805*, ed. Metche Budka (Elizabeth, NJ: Grassman Publishing Co., 1965), 93.

[11] Gibbs Myers, "Pioneers of the Federal Area," *Records of the Columbia Historical Society*, vol. 44–45 (Washington: Columbia Historical Society, 1944), 130.

[12] Myers 147.

[13] Arnebeck 263.

[14] Anne Newport Royall, *Sketches of History, Life, and Manners, in the United States* (1826; New York: Johnson Reprint Corp., 1970), 174.

[15] *Records of the District of Columbia Commissioners*, Proceedings, 14 March 1793, NA, Roll 1, p. 173.

Notes

[16] H. Paul Caemmerer, *The Life of Pierre Charles L'Enfant* (Washington: National Republic P, 1950), p. 187.

[17] *Records of the District of Columbia Commissioners*, Proceedings, 28 March 1792, NA.

[18] Tindall 180.

[19] Niemcewicz 80–81.

[20] Record Group 217, Miscellaneous Treasury Records, NA, Box, no. 48, Time notes for the stone cutters at the President's House for December 1794.

[21] Heaton, Ronald E. and James R. Case, *The Lodge at Fredericksburg, A Digest of the Early Records* (Bloomington: Pantagraph P, 1975), 27–31.

[22] Glen Brown, *History of the United States Capitol*, vol. l, (Washington: U.S. GPO, 1900), 100–01.

[23] Tindall 180

[24] Account book of Edrington and Moncure Quarry, 1836–1839, July 1838, Virginia Historical Society Library, Richmond.

[25] Edrington and Moncure Account Book, 1839, 38.

[26] Tindall 182.

[27] *Records of the District of Columbia Commissioners*, Proceedings, 1793, NA, roll 1, 177.

[28] *Records of the District of Columbia Commissioners*, Proceedings, 29 July-3 August 1793, NA, Roll 1, 192.

[29] Bob Arnebeck, *Through a Fiery Trial: Building Washington 1790–1800* (Lanham, MD: Madison Books, 1991), 232.

[30] William Seale, *The President's House*, vol. 1 (Washington: White House Historical Association—National Geographic Society, 1986), 67.

[31] Fredericksburg, Virginia, Court Records, Box 51, Exhibit G, Cook vs. Brent. List of Debts by John Cooke, 2 January 1809.

[32] Edrington and Moncure Account Book, 6 November 1840, 102.

[33] Frank E. Edgington, *A History of the New York Avenue Presbyterian Church* (Washington: New York Avenue Presbyterian Church, 1961), 3.

[34] Seale, vol. 1, 67.

[35] Constance McLaughlin Green, *Washington Village and Capital 1800–1878* (Princeton: Princeton UP, 1962), 37–38.

Notes

8. Quarrying Aqua Stone

[1] Harley J. McKee, *Introduction to Early American Masonry* (Washington: National Trust for Historic Preservation, 1980), 24.

[2] Patrick Plunkett, personal interview, 5 August 1993.

[3] Lee Nelson, *White House Stone Caring: Builders and Restorers* (Washington D.C., U.S. GPO, 1992), 4

[4] *Records of the District of Columbia Commissioners*, Proceedings, 30 March 1792, NA, M-371, Roll 1, p. 91.

[5] McKee 22–24.

[6] Nelson 6.

[7] *Records of the District of Columbia Commissioners*, Letters received, Samuel Milliken to the Commissioners, 27 October 1792, NA, roll 9, no. 145.

[8] *Thomas Jefferson and the National Capital*, ed. Saul K. Padover (Washington, D.C.: U.S. GPO, 1946), 157.

[9] *Thomas Jefferson and the National Capital*, 162–63.

[10] McKee 18.

[11] Wilbur Segar, personal interview, 5 July 1994.

9. Transporting the Stone

[1] Land Survey Maps 1877 and 1913, Commissioner of Revenue's Files, Stafford Administration Building, Stafford, Virginia.

[2] Milton Dickerson, personal interview, 12 February 1994.

[3] *Records of the District of Columbia Commissioners*, Letters received. Thomas Towson to Commissioner of Public Buildings, 11 March 1824, NA, M–371.

[4] *Records of the District of Columbia Commissioners*, Letters sent, Commissioners to John Gibson, 2 January 1798, NA.

[5] Benjamin Henry Latrobe, *The Correspondence and Miscellaneous Papers of Benjamin Henry Latrobe*, Series 4, vol. 2. Ed. John C. Van Horne. Written by Latrobe 28 November 1806. (New Haven: Yale UP, 1986), 299–300.

[6] George Gordon, personal interview, 11 February 1994.

[7] *Records of the District of Columbia Commissioners*, Proceedings, 1 May 1792, NA, Roll 1, 103.

[8] Record Group 42, NA, Box l, no. 109.

Notes

[9] Bob Arnebeck, *Through a Fiery Trial; Building Washington 1790–1800* (Lanham: Madison Books, 1991), 248.

[10] Arnebeck 232.

[11] *Records of the District of Columbia Commissions*, Letters sent, Joseph Elgar to Withers Waller, 17 January 1824, NA.

[12] *Records of the District of Columbia Commissioners*, Proceedings, 4 June 1792, NA, roll 1, 111.

[13] Arnebeck 233.

[14] *Records of the District of Columbia Commissioners*, Letters received, O'Neale to Commissioners, 19 September 1794, NA.

[15] Arnebeck 316.

[16] *An Illustrated History: The City of Washington Junior League of Washington*, ed. Thomas Froncek (New York: Wing Books, 1977), 122.

10. Work Continues

[1] *Records of the District of Columbia Commissioners*, Proceedings, 10 April 1972, NA, Roll 1, p. 93.

[2] *Records of the District of Columbia Commissioners*, Letters received, Wright to Commissioners, December 1792, NA.

[3] *Records of the District of Columbia Commissioners*, Proceedings, 1 January 1793, NA, M-371.

[4] *Records of the District of Columbia Commissioners*, Proceedings, 15 October 1793, NA, M-371.

[5] Bob Arnebeck, *Through A Fiery Trial: Building Washington 1790–1800* (Lanham: Madison Books, 1991), 206.

[6] *Records of the District of Columbia Commissioners*, Letters received. Hoban to Commissioners, 19 November 1793, NA.

[7] *Records of the District of Columbia Commissioners*, Proceedings, August 1793, NA, M-371.

[8] I. T. Frary, *They Built the Capitol* (Richmond: Garrett and Massie, 1940), 161–62.

[9] Glenn Brown, *History of the United States Capitol*, vol. 1 (Washington: U.S. GPO, 1900), 61.

[10] Arnebeck 246.

Notes

[11] Arnebeck 218.

[12] *Records of the District of Columbia Commissioners*, Letters received, Masons to Commissioners, 23 June 1794, NA.

[13] Arnebeck 290.

[14] Arnebeck 254.

[15] Arnebeck 301.

[16] George S. Hunsberger, "The Architectural Career of George Hadfield." *Records of the Columbia Historical Society*, vol. 51–52 (Washington: Columbia Historical Society, 1955), 46–65.

[17] *Records of the District of Columbia Commissioners*, Letters received. George Blagden to the Commissioners, 6 August 1975, NA.

[18] *Records of the District of Columbia Commissioners*, Letters received. George Blagden to the Commissioners, 19 October 1975, NA.

[19] Arnebeck 386.

[20] *Records of the District of Columbia Commissioners*, Proceedings. 11 October 1796 and 24 October 1796, NA, vol. 3: 208, 211.

[21] *Records of the District of Columbia Commissioners*, Letters received. Stone Carvers to Commissioners, 17 March 1797, NA. Proceedings, 22 March, 3 April, and 3 May 1797.

[22] Collen Williamson, letter to President Adams, 26 April 1797, Adams Family Papers, LOC, Reel 384.

[23] Glenn Brown, *History of the United States Capitol* vol. 1 (Washington: U.S. GPO, 1900), 61.

[24] Arnebeck 472.

[25] *Records of the District of Columbia Commissioners*, Proceedings. 22 February 1798, NA, vol. 4: 236.

[26] *Records of the District of Columbia Commissioners*, Letters received, Capitol stone carvers to Commissioners, 16 April 1798, NA.

[27] *Records of the District of Columbia Commissioners*, Letters received, Capitol stone cutters to Commissioners, 16 April 1798, NA.

[28] *Records of the District of Columbia Commissioners*, Letters sent, Commissioners to the stone cutters at both sites, 16 April 1798, NA, roll 4, 979.

[29] *Records of the District of Columbia Commissioners*, Letters sent, Commissioner to Robert Stewart, 18 April 1798, NA, roll 4, 983.

[30] *Records of the District of Columbia Commissioners*, Letters received, George Hadfield to the Commissioners, 30 April 1798, NA, roll 13, 1347.

[31] George S. Hunsberger, "The Architectural Career of George Hadfield." *Records of the Columbia Historical Society*, vols. 51–52 (Washington: Columbia Historical Society, 1955), 46–65.

[32] Frary 45.

[33] Julian Ursyn Niemcewicz, *Under Their Vine and Fig Tree: Travels Through America 1797–1799, 1805*, ed. Metche Budka (Elizabeth, N.J.: Grassman Publishing Co., 1965), 78.

[34] Wilhelmus Bogart Bryan, *A History of the National Capital*, vol. 1, (New York: MacMillan Company, 1914), 313.

11. Moving into the Federal Buildings

[1] John Ball Osborne, "The Removal of the Government to Washington," *Records of the Columbia Historical Society*, vol. 3 (Washington: Columbia Historical Society, 1900), 139.

[2] Abigail Adams, *New Letters of Abigail Adams, 1788–1801*, ed. Stewart Mitchell (Boston: Houghton Mifflin Co., 1947), 257–59.

[3] Anna Maria Brodeau Thornton, "Diary of Mrs. William Thornton, 1800–1863," *Records of the Columbia Historical Society* (Washington: Columbia Historical Society, 1907), 119–120.

[4] Glenn Brown, *History of the United States Capitol*, vol. 2 (Washington: U.S. GPO, 1903), 129.

[5] *The State of the Union Messages of the Presidents 1790–1860*, ed. Fred L. Israel, vol. 1. State of Union message delivered on 22 November 1800 by John Adams. (New York: Chelsea House P, 1966), 52.

[6] Brown 134.

12. Benjamin Henry Latrobe's Contributions

[1] Benjamin Henry Latrobe, *The Journals of Benjamin Henry Latrobe: 1799–1820, Philadelphia to New Orleans*, ed. Edward C. Carter II, John C. Van Horne, and Lee W. Formwalt (New Haven: Yale UP for Maryland Historical Society, 1980), 91.

[2] Latrobe, Journals 92.

[3] Benjamin Henry Latrobe, *Latrobe's View of America, 1795–1820: Selections from the Watercolors and Sketches*, ed. Edward C. Carter II, John C. Van Horne, and Charles E. Brownell (New Haven: Yale UP for Maryland Historical Society, 1985), 272.

Notes

[4] Latrobe, Journals 96.

[5] Latrobe, Journals 93.

[6] Latrobe, Journals 78–80.

[7] Latrobe, Journals 65ff.

[8] Latrobe, Journals 65.

[9] Latrobe, Journals 69–71.

13. Other Freestone Quarries

[1] *Records of the District of Columbia Commissioners*, Letters sent, Commissioners to John Gibson, 2 January 1796, NA, M-371.

[2] *Virginia Herald*, Fredericksburg, Virginia, 22 December 1794.

[3] Benjamin Henry Latrobe, *The Papers of Benjamin Henry Latrobe; Correspondence and Miscellaneous Papers*, Series 4, vol. 1, ed. John C. Van Horne and Lee W. Formwalt (New Haven: Yale UP, 1984), 426n.

[4] Latrobe 425.

[5] George Washington, letter to Daniel Carroll, 16 December 1793, *Writings of George Washington*, ed. John C. Fitzpatrick, vol. 33 (Washington: U.S. GPO, 1937), 184.

[6] Bob Arnebeck, *Through a Fiery Trial: Building Washington 1790–1800* (Lanham: Madison Books, 1991), 214.

[7] George Washington, letter to William Pearce, 21 September 1794, *Writings*, vol. 33, 502.

[8] George Washington, letter to William Pearce, 27 July 1794, *Writings*, vol. 33, 447.

[9] George Washington, letter to William Pearce, 14 December 1794, *Writings*, vol. 34, 59.

[10] George Washington, letter to William Pearce, 26 October 1796, *Writings*, vol. 35, 251.

[11] William Pearce, letter to George Washington, 13 November 1796, Notebook 16, NA.

[12] *Records of the District of Columbia Commissioners*, letters received, John Dunbar to the Commissioners, 18 May 1795 and 21 September 1795, NA, M-371, roll 8, 581 and 662.

[13] *Impartial Observer and Washington Advertiser*, 31 July 1795.

[14] *Records of the District of Columbia Commissioners*, letters received, John Dunbar to

the Commissioners from Dumfries, Virginia, jail, 5 May 1796 and 9 May 1796, nos. 1115 and 118 NA.

[15] *The Columbian Chronicle*, Georgetown, 24 February 1796, 1.

[16] Latrobe 425.

[17] Latrobe 427n.

[18] Latrobe 425.

[19] *Washington Gazette*, 14 October 1797.

[20] *Centinel of Liberty*, 29 March 1799.

[21] Latrobe 427n.

[22] Information for notes obtained from Barbara Schomp Kirby, Stafford County, Virginia.

14. Cutting and Carving Aqua Stone

[1] Lee H. Nelson, *White House Stone Carving: Builders and Restorers* (Washington: U.S. GPO, 1992), 10.

[2] Harley J. McKee, *Introduction to Early American Masonry: Stone, Brick, Mortar and Plaster* (Washington: National Trust for Historic Preservation, Preservation P, 1980), 29.

[3] Nelson 12.

[4] Nelson 10.

[5] McKee 31–32.

[6] Nelson 12.

[7] McKee 62–63.

[8] Patrick J. Plunkett, personal interview, 5 August 1993.

[9] Anna Maria Brodeau Thornton, "Diary of Mrs. William Thornton, 1800–1863," *Records of the Columbia Historical Society* vol. 10 (Washington: Columbia Historical Society, 1907), 119–120.

[10] Nelson 16, 18, 21.

[11] I. T. Frary, *They Built the Capitol* (Richmond: Garrett and Massie, 1940), 107–8.

[12] Benjamin Henry Latrobe, *The Correspondence and Miscellaneous Papers of Benjamin Henry Latrobe*, vol. 3, ed. John C. Van Horne (New Haven: Yale UP, 1988), 26.

[13] Frary 109.

[14] Frary 109–10.

[15] Talbot Hamlin, *Benjamin Henry Latrobe* (New York: Oxford UP, 1955), 268.

Notes

16 Hamlin 270.

17 Latrobe 328.

18 Frary 123–125.

19 Frary 121, 123.

20 John Quincy Adams, *The Diary of John Quincy Adams*, ed. Allan Nevins, 1951), 366.

21 Glenn Brown, *History of the United States Capitol*, vol. 2 (Washington: GPO, 1903), 183.

22 Glenn Brown, *History of the United States Capitol*; The Old Capitol—1792–1850, vol. 1 (Washington: GPO, 1900), 75.

23 Brown, vol. 2, 183.

24 Latrobe 27.

25 Frary 116.

26 Robert James Kapsch, "The Labor History of the Construction and Reconstruction of the White House, 1791–1817," diss., U of Maryland, 1992, 227n.

27 Kapsch 227–28n.

15. War of 1812

1 *Funk & Wagnalls New Encyclopedia*, ed. Robert S. Phillips, 4th ed., vol. 27 (New York: Funk & Wagnalls, 1983), 143.

2 Dolley Madison, *Memoirs and Letters of Dolly [sic] Madison*, ed. Lucia Beverly Cutts (Port Washington, NY: Kennikat P, 1971), 90–91.

3 Madison 108–109.

4 Madison 110–111.

5 Madison 111.

6 Lonnelle Aikman, *We the People*, 13th. ed. (Washington: U.S. Capitol Historical Society with National Geographic Society, 1985), 28–29, 31.

7 Irving Brant, *James Madison, Commander in Chief: 1812–1836* (Indianapolis: Bobbs-Merrill Co., Inc., 1961), 305.

8 William Seale, *The President's House*, vol. 1 (Washington: White House Historical Society—National Geographic Society, 1986), 135.

9 Donald R. Hickey, *The War of 1812: A Forgotten Conflict* (Urbana: U of Illinois P, 1989), 199.

10 Margaret Bayard Smith, *The First Forty Years of Washington Society*, ed. Gaillard

Hunt, 2nd ed. (New York: Frederick Ungar P 1965), 111–12.

[11] Hickey 199.

[12] Brant 304.

[13] Madison 113.

[14] Hickey 199.

[15] Hickey 202.

[16] Smith 109.

[17] Hickey 202.

[18] Hickey 235.

[19] Brant 323.

[20] Emmie Ferguson Farrar and Emilee Hines, *Old Virginia Houses: The Northern Peninsula* (Charlotte: Delmar Co., 1972), 51.

Edith Tunis Sale, *Interiors of Virginia Houses of Colonial Times* (Richmond: William Bryd P, Inc., 1927), 151.

[21] Hickey 135.

[22] Hickey 119.

[23] John Quincy Adams, The Writings of John Quincy Adams, ed. Worthington Chauncey Ford (New York: MacMillan Co., 1915), 149.

[24] Smith 115.

[25] Hickey 241.

[26] George McCue, *The Octagon* (Washington: American Institute of Architects Foundation, 1976), 62.

16. Rebuilding

[1] *Records of the District of Columbia Commissioners*, Letters received, Hadfield to Commissioners, 13 October 1814, NA, M-371, roll 18, 2162.

[2] William Seale, *The President's House,* vol. I (Washington: Washington Historical Association—National Geographic Society, 1961), 139.

[3] Donald R. Hickey, *The War of 1812: A Forgotten Conflict* (Urbana: U of Illinois P., 1989), 235.

[4] Constance McLaughlin Green, Washington Village and Capital, 1800–1878 (Princeton, NJ: Princeton U P, 1962), 67.

[5] Benjamin Henry Latrobe, *The Correspondence and Miscellaneous Papers of*

Notes

Benjamin Henry Latrobe, ed. John C. Van Horne, vol. 3 (New Haven: Yale UP, 1988), 644.

[6] Latrobe 670.

[7] Latrobe 650.

[8] Latrobe 651.

[9] Latrobe 653.

[10] Latrobe 661.

[11] Latrobe 681–82.

[12] Latrobe 683.

[13] Seale 139.

[14] *Records of the District of Columbia Commissioners*, Letters received, Hoban to the Commissioners, 25 April 1815, NA, roll. 18, 2188.

[15] Robert James Kapsch, "The Labor History of the Construction and Reconstruction of the White House, 1791–1817," diss., U of Maryland, 1992, 325.

[16] Notice sent by Office of Commissioners of the Public Buildings on 10 February 1816 to newspapers. *Daily National Intelligencer*, Washington, 10 February 1816: 3.

[17] Commissioners sent advertisement 1 April 1816 to newspapers along East Coast. *Daily National Intelligencer*, Washington, 4 April 1816: 1.

[18] *Records of the District of Columbia Commissioners*, Letters sent, Commissioners to John Brannan, agent in Baltimore, 28 August 1815, NA, roll 5, 80–81.

[19] *Records of the District of Columbia Commissioners*, Letters sent, Commissioners to Capitol stone cutters, 27 October 1815, NA, roll 5, 102–04.

[20] Glenn Brown, *History of the United States Capitol*, vol. 1 (Washington: GPO, 1900), 101.

[21] Talbot Hamlin, *Benjamin Henry Latrobe, 1799–1820*, (New York: Oxford UP, 1955), 448.

[22] Seale 141.

[23] Latrobe 852.

[24] Latrobe 921.

[25] Hamlin 270.

[26] *Records of the District of Columbia Commissioners*, Letters sent, Commissioners to Hoban and Latrobe, 4 April 1816, NA, roll 5, 147.

[27] *Records of the District of Columbia Commissioners*, Letters received, Latrobe to Lane, 1 May 1816, NA, Roll 19, 2309.

[28] *Records of the District of Columbia Commissioners*, Letters received, Lennox to Lane, 22 October 1816, NA.

[29] Seale 146.

[30] Hamlin 477.

[31] I. T. Frary, *They Built the Capitol* (Richmond: Garrett and Massie, 1940), 133–41.

[32] Lonnelle Aikman, *We the People*, 13th ed. (Washington: U.S. Capitol Historical Society with National Geographic Society, 1985), 35.

[33] William C. Allen, *The Dome of the United States Capitol: An Architectural History* (Washington, U.S. GPO, 1992), 5.

[34] Anne Newport Royall, *Sketches of History, Life and Manners in the United States* (1826; New York: Johnson Reprint Corp., 1970), 134.

17. Capitol's Completion and Quarries' Closure

[1] Bob Arnebeck, *Through A Fiery Trial* (Lanham, MD: Madison Books, 1991), 464.

[2] *Funk & Wagnalls New Encyclopedia*, ed. Robert S. Phillips, 4th ed, vol. 7 (New York: Funk & Wagnalls, 1983), 31.

[3] William C. Allen, *Capitol Columns at the National Arboretum: An Architectural History,* prepared under direction of George M. White, FAIA Architect of the Capitol, 1989 6–7.

[4] *Records of the District of Columbia Commissioners*, Letters received, Thomas Towson, quarrier, to Joseph Elgar, Commissioner of Public Buildings, 7 October 1823 NA, Roll 21, 2702.

[5] *Records of the District of Columbia Commissioners*, Letters received, Towson to Elgar, 30 October 1823 NA. roll 21, 2702.

[6] *Records of the District of Columbia Commissioners*, Letters received, Towson to Elgar, 5 November 1823 NA, roll 21, 2702.

[7] *Records of the District of Columbia Commissioners*, Letters sent, Elgar to Waller, 17 January 1824 NA, 351.

[8] *Records of the District of Columbia Commissioners*, Letters sent, Elgar to Waller and Morton, 13 February 1824 NA, 352.

[9] *Records of the District of Columbia Commissioners*, Letters sent, Elgar to Towson, 13 February 1824 NA, 352.

[10] *Records of the District of Columbia Commissioners*, Letters received, Towson to Elgar, 11 March 1824 NA, roll 21, 2714.

[11] "The Stonecutters Celebration," *Washington Gazette*, 2 July 1824.

Notes

[12] "Celebration of the Fourth of July," *Washington Gazette*, 6 July 1824.

[13] *Letters of the District of Columbia Commissioners*, Letters received, George Blagden to Joseph Elgar, 21 April 1824, NA 27ll.

[14] *Letters of the District of Columbia Commissioners*, Letters received, William Stewart to Elgar, 27 July 1824, NA, 2726.

[15] *Letters of the District of Columbia Commissioners*, Letters received, Towson to Elgar, 29 March 1825, NA, Roll 21, 2733.

[16] *The National Intelligencer*, 19 May 1824.

[17] Anne Newport Royall, *Sketches of History, Life, and Manners in the United States* (1826; New York: Johnson Reprint Corp., 1970), 175.

[18] Benjamin Henry Latrobe, *The Papers of Benjamin Henry Latrobe; Correspondence and Miscellaneous Papers*, vol. 3, ed. John C. Van Horne (New Haven: Yale UP for Maryland Historical Society, 1988), 328.

[19] Allen 12.

[20] Allen 13.

[21] *Stafford Co., Virginia 1800–1850*, Census compiled by A. Maxim Coppage and James Wm. Tackitt (Concord, Calif.: n.p. 1982), 30.

[22] *Executive Papers*, Governor John Floyd from George M. Cooke, 13 September 1831, Archives Branch, Virginia State Library.

[23] H. Paul Caemmerer, "The Sesquicentennial of the Laying of the Cornerstone of the United States Capitol by George Washington," *Records of the Columbia Historical Society* vol. 44–45, (Washington: Columbia Historical Society, 1944), 184.

[24] Robert Dale Owen, *Hints on Public Architecture* (Washington, 1849), 113.

[25] Lonelle Aikman, *We the People*, 13th ed. (Washington: U.S. Capitol Historical Society with National Geographic Society, 1985), 41.

[26] Aikman 51.

[27] Aikman 43.

[28] William C. Allen, personal interview, 18 August 1993.

[29] Allen, *Dome* 11.

[30] Allen, *Dome* 13.

[31] Allen, *Dome* 15.

[32] *Documentary History of the Construction and Development of the United States*

Capitol Buildings and Grounds (Washington: U.S. GPO, 1904), 1018.

[33] Allen, *Dome* 27ff.

[34] *Documentary History* 1018.

[35] Memorandum from Joseph F. Robinson to J. George Stewart, 5 May 1959, Architect of the Capitol's Files.

[36] Lawrence L. Knutson, "Coming Down for Cleanup," *Free Lance-Star*, Fredericksburg, 16 April 1993, 1A.

[37] John T. Goolrick, *The Story of Stafford* (Stafford: Cardinal Press, 1988), 75–77.

[38] Margaret Waller Ford, letter to Evelyn Lewis, 1882. Aquia Harbour Historical Files, Stafford, Virginia.

[39] George Gordon, personal interview, 11 February 1994.

[40] Jack Patterson, "Aquia: An Illustrious Past," *Aquia Quarterly*, Aquia Harbour, vol. 4, no. 1, spring 1979, 14.

[41] Goolrick 7844

[42] Untitled booklet from George Washington Stone Corporation. Probably printed in 1920s. Copy in Architect of Capitol's files and at the Virginia Historical Library, Richmond, Virginia, 9–12.

18. Truman's White House Renovation

[1] Harry S. Truman, *Off the Record: The Private Papers of Harry S. Truman*, ed. Robert H. Ferrell (New York: Harper & Row, 1980), 242–43.

[2] Margaret Truman (with Margaret Cousins), *Souvenir. Margaret Truman's Own Story* (New York: McGraw-Hill, 1956), 246.

[3] Harry S. Truman, letter to Mary Jane Truman, 10 Aug. 1948, Harry S. Truman Library, Independence, Mo.

[4] *Report of the Commission on the Renovation of the Executive Mansion*, compiled by Edwin Bateman Morris (Washington: U.S. GPO, 1952), 39.

[5] David McCullough, *Truman* (New York: Simon & Schuster, 1992), 594.

[6] *Report of Comm.* 39.

[7] *Report of Comm.* 99.

[8] Harry S. Truman, letter to Clarence Cannon, Washington, 3 May 1949, Harry S. Truman Library, Independence, Mo.

[9] McCullough 877.

Notes

[10] *Report of Comm.* 49, 51.

[11] Mount Vernon's *Superintendent's Monthly Report*, March 1954, 1.

[12] *Report of Comm.* 104.

[13] *Report of Comm.* 20 June 1950. 94.

[14] Rex Scouten, personal interview, 15 October 1993.

[15] Margaret Truman 310.

[16] S. Brent Morris, *Cornerstones of Freedom: A Masonic Tradition*, ed. John W. Boettjer. Letter sent from Truman to Masonic Grand Lodges, 22 November 1952 (Washington: The Supreme Council, 33^0, S.J.,1993), 82.

[17] Truman, *Off the Record* 243.

[18] *Report of Comm.* 105.

[19] Truman, *Off the Record* 246.

[20] McCullough 886.

[21] Seale 1027.

19. Extending the Capitol's East Front

[1] Lonnelle Aikman, *We the People* 13th. ed. (Washington; U.S. Capitol Historical Society with National Geographic Society, 1985), 57–58.

[2] "White House Tour: Music by Truman," *New York Times*, 4 May 1952: 53.

[3] Harry S. Truman, *Off the Record: The Private Papers of Harry S. Truman*, ed. Robert H. Ferrell (New York: Harper & Row, 1980), 357–58.

[4] Aikman 57.

[5] Memorandum, Office of the Curator

[6] Phillip Morris, "Our Capitol-in-a-Meadow," *Southern Living*, vol. 27, March 1992: 66.

[7] William C. Allen, letter to author, 1993.

[8] Reed Black, personal interview, 11 August 1994.

20. Who Owned Government Island?

[1] Deed from Governor F. W. M. Holliday to C. A. Bryan and S. B. Howell, 1 April 1879, Stafford County Clerk's Office, Stafford County Courthouse, Stafford, Virginia.

[2] Marvin C. Bowling, "Who Owned Government Island?" *Lawyers Title News*, February 1964: 5–7.

[3] William Lakeman, "Proposed Park Awaits Study of Ownership," *Free Lance-Star*, 21 February 1963.

Notes

⁴ Thomas Metts, personal interview, 13 July 1993.

⁵ "Commuter Heads for His Potomac 'Isle,'" *Washington Post*, 18 May 1964, D1.

21. Deterioration of Aquia Stone

¹ Harley J. McKee, *Introduction to Early American Masonry: Stone, Brick, Mortar and Plaster* (Washington: National Trust for Historic Preservation, Preservation P, 1980), 33.

² *The Old Cape Henry Lighthouse* Historic Structures Report Phase II, prepared for the Society for the Preservation of Virginia. Antiquities by Wood, Sweet, and Swofford Architects, Charlottesville, 1991: 6.

³ *Cape Henry* 15.

⁴ *Cape Henry* 4.

⁵ *Cape Henry* 15.

⁶ *Cape Henry* 4.

⁷ William C. Allen, "The Capitol's Four Cornerstones," address, Bicentennial of the Laying of the Cornerstone of the U.S. Capitol, U.S. Capitol Historical Society Conference, 18 September 1993.

⁸ William C. Allen, personal interview, 18 September 1993.

⁹ Rex Scouten, personal interview, 20 August 1992.

22. Painting and Restoring the Stone of the White House

¹ Wilhelmus Bogart Bryan, *A History of the National Capitol* (New York: MacMillan Company, 1914), 313.

² National Archives RG 217, Accounts of the Commission of the City of Washington 1794–1800, Roll of Painters at President's House, 1798. Box 5, Voucher 77. Hoban lists names: Lewis Clephan, John D. Lowrey, and Thomas Money.

³ *Mining Journal* (London: 1849). Reprinted in *Chambers' Journal* (London: 1849), *Decoration*, an English magazine, vol. 3 (January/June 1882). At Avery Architecture Library, Columbia University, New York.

⁴ *The History of the White House* (London: Odyssey Publishing & Mallard Press, 1991), 1.

⁵ Wilhelmus Bogart Bryan, "The Name White House," *Records of the Columbia Historical Society*, vol. 33–34 (Washington, D.C., Columbia Historical Society, 1932), 308.

⁶ *The History of the White House*, 1.

Notes

[7] Thomas H. Rudder, letter to Gary J. Walters, 18 October 1991, Files of White House Chief Usher.

[8] Rex Scouten, personal interview, 5 August 1993.

[9] James I. McDaniel, "Stone Walls Preserved," *White House History Journal*, vol. 1 no. 1 (Washington: White House Historical Association, 1983), 40–41.

[10] Rex Scouten, personal interview, 5 August 1993.

[11] Jimmy Carter, note to author, 10 October 1993.

[12] Rudder, letter to Walters.

[13] Rex Scouten, personal interview, 5 August 1993.

[14] Robert James Kapsch, "The Labor History of the Construction and Reconstruction of the White House, 1791–1817," diss., U of Maryland, 1992, 33.

[15] McDaniel 42.

[16] Lee H. Nelson, *White House Stone Carving* (Washington: U.S. GPO, 1992), 25–26.

[17] W. Dale Nelson, "Presidential Face Lift," *Free Lance-Star*, Fredericksburg, Virginia, 15 May 1989: 15.

[18] Patrick J. Plunkett, personal interview, 20 September 1994.

[19] Nelson 26.

[20] George H. W. Bush, note to Gary Walters, 15 June 1987. Files of White House chief usher.

[21] Patrick J. Plunkett, personal interview, 5 August 1993.

[22] George H. W. Bush, phone message to author, 2 September 1993.

[23] Ronald Reagan, memorandum to author, 20 September 1993.

[24] Rudder, letter to Walters.

23. Removing and Replacing the Capitol's Stone

[1] Drew Pearson, "The Crumbling Capitol," *Washington Post*, 11 July 1965.

[2] Congressional Record, 88 Congress 2nd Sess., 17 September 1965.

[3] William Allen, *History of the United States Capitol* (Washington: U.S. GPO 2001), 442.

[4] Edward O'Neil, *Repair and Maintenance of Masonry Structures: Case Histories*, (Washington, D.C., Army Corps of Engineers, 1995), 8–9.

[5] William Allen, personal interview, 18 August 1993.

[6] Rudder, letter to Walters.

[7] Brochure, West Front Renovation, Architect of the Capitol's Office.

Epilogue

[1] Margaret Bayard Smith, *The First Forty Years of Washington Society*, ed. Gillard Hunt (New York: Frederick Ungar Press,1968), 383.

Appendix II: Other Structures of Aquia Stone

[1] Henry Irving Brock, *Colonial Churches in Virginia* (Port Washington, NY: 1930), 4.

[2] Thomas Tileston Waterman, *Mansions of Virginia* (Chapel Hill: U of North Carolina P, 1945), 417.

[3] George Mason, *Papers of George Mason*, ed. Rutland. Letter to John Mason, 20 August 1792, 1272.

[4] Mrs. Henry Gwynne Tayloe, Jr., personal interview, 4 January 1994.

[5] George McCue, *The Octagon*, (Washington, D.C.: American Institute of Architects Foundation, 1976), 18, 81.

[6] Waterman 30.

[7] *Calendar of State Papers*, vol. 5, Letter written to Governor of Virginia on 18 July 1791, 343.

[8] Charles C. Brownell, *Making of Virginia Architecture*, Virginia Museum of Fine Arts (Charlottesville: UP, 1992), 222.

[9] John William Reps, *Washington on View: The Nation's Capital Since 1790* (Chapel Hill: U of NC P, 1991), 25.

[10] Harlan D. Unrau, "Chronological History of the Chesapeake and Ohio Canal; 1828–1924," a Historical Resource Study, vol. 4, Seneca, Md., May 1976.

[11] Phyllis Sprock, personal interview, 30 September 1994. Fort Monroe, Chief of Environmental Office.

[12] William Seale, *The President's House*, vol. 2 (Washington, D.C.: White Historical Association, 1986), 197.

[13] *Historic American Buildings Survey*, National Park Service, Department of the Interior, *District of Columbia Catalog*, compiled by Nancy B. Schwartz (Charlottesville: UP of Virginia, 1974), 111–12.

[14] *The City of Washington; An Illustrated History* by the Junior League of Washington, ed. Thomas Froncek (New York: Knopf, 1977), 182.

[15] Mary Louise Thompson, personal interview, 28 September 1994. Washington National Cathedral.

Bibliography

Adams, Abigail. *New Letters of Abigail Adams 1788–1801*. Ed. Stewart Mitchell. Boston: Houghton Mifflin Co., 1947.

Adams, John Quincy. T*he Diary of John Quincy Adams, 1794–1845*. Ed. Alan Nevins. New York: Scribner, 1951.

Aikman, Lonnelle. *We, the People*. 13th ed. Washington, D.C.: U.S. Capitol Historical Society with National Geographical Society, 1985.

Allen, Williams C. *Capitol Columns at the National Arboretum*. Washington, D.C. Prepared under the direction of George M. White, FAIA, Architect of the Capitol, 1989.

——. *History of the United States Capitol: A Chronicle of Design, Construction and Politics*. Washington, D.C.: U.S. Government Printing Office, 2001.

——. *The Dome of the United States Capitol; An Architectural History*. Washington, D.C.: U.S. Government Printing Office, 1992.

——. *The United States Capitol: A Brief Architectural History*. Washington, D.C.: U.S. Government Printing Office, 1990.

Arnebeck, Bob. *Through a Fiery Trial: Building Washington, 1790–1800*. Lanham: Madison Books, 1991.

Bedini, Silvio. *The Life of Benjamin Banneker*. Rancho Cordova, CA: Landmark Enterprises, 1972.

Boundary Markers of the Nation's Capital. A National Capital Planning Commission Bicentennial Report. Washington, D.C.: Government Printing Office. Summer, 1976.

Bowling, Kenneth R. *The Creation of Washington, D.C. The Idea and Location of the American Capital*. Fairfax, VA: George Mason University Press, 1991.

——. "The Other G.W." *Washington History*. Vol. 3, no. 2, Fall/Winter 1991–92.

——. *Peter Charles L'Enfant, Vision, Honor and Male Friendship in the Early American Republic*. Washington: Friends of the GWU Libraries, 2002.

Brant, Irving. *James Madison, Commander in Chief: 1812–1836*. Indianapolis: Bobbs-Merrill Co., Inc., 1961.

Brock, Henry Irving. *Colonial Churches in Virginia*. Port Washington, D.C.: Dale Press, 1930.

Brown, Glenn. *History of the United States Capitol*. 2 vols. Washington, D.C.: U.S. Government Printing Office, 1900 and 1903.

Brown, Glenn. "The United States Capitol in 1800." *Records of the Columbia Historical Society*. Vol. 4. Lancaster, PA: Published by the Society, New Era Printing Co., 1901. 128–135.

Brownell, Charles C. [*et al*] *Making of Virginia Architecture*. Virginia Museum of Fine Arts. Charlottesville: University Press of Virginia, 1992.

Bryan, Wilhelmus Bogart. *A History of the National Capital*. 2 vols. New York: Macmillan, 1914 and 1916.

Building Stones of our Nation's Capital. U.S. Dept. of the Interior Geological Survey. Washington, D.C.: U.S. Government Printing Office, 1975.

Caemmerer, H. Paul. *The Life of Pierre Charles L'Enfant, Planner of the City Beautiful, The City of Washington*. Washington, D.C.: National Republic Press, 1950.

Bibliography

Callahan, Charles. *Washington the Man and the Mason.* Washington, D.C.: Gibson Bros., 1913.

———. "The Sesquicentennial of the Laying of the Cornerstone of the United States Capitol by George Washington." *Records of the Columbia Historical Society.* Vol. 44–45. Washington, D.C.: Columbia Historical Society, 1944. 161–89.

Calendar of State Papers. Richmond: 1885. Vol. 5. New York: Kraus Reprint Corp., 1968.

Clark, Appleton, P. "Origin of the Building Regulations." *Records of the Columbia Historical Society.* Vol. 4, Baltimore: New Era Printing Co., 1901. 166–173.

Curl, James Stevens. *The Art and Architecture of Freemasonry, An Introductory Study.* London: B. T. Batsford, Ltd., 1991.

Cutts, J. Madison. "Dolly Madison." *Records of the Columbia Historical Society.* Vol. 3. Published by the Society, 1901. 28–72.

Documentary History of the Construction and Development of the United States Capitol Buildings and Grounds. Washington, D.C.: U.S. Government Printing Office, 1904.

Edgington, Frank E. *A History of the New York Avenue Presbyterian Church.* Washington, D.C.: New York Avenue Presbyterian Church, 1961.

Farrar, Emmie Ferguson and Emilee Hines. *Old Virginia Houses: The Northern Peninsula.* Charlotte: The Delmar Co., 1972.

Ferris, Robert. *Founders and Frontiersmen.* Washington, D.C.: U.S. Government Printing Office, 1967.

Fitzgerald, Ruth Coder. *A Different Story.* Fredericksburg, VA: Unicorn, 1979.

Ford, Worthington Chaucey, ed. *Writings of John Quincy Adams 1814–1816.* Vol. 5. New York: The Macmillan Co., 1915.

Frary, I. T. *They Built the Capitol.* Richmond: Garrett and Massie, 1940.

Froncek, Thomas, ed. *An Illustrated History: The City of Washington.* Junior League of Washington. New York: Wing Books, 1977.

Funk & Wagnalls New Encyclopedia. Ed. Robert S. Phillips, 4th ed., vols. 7, 27. New York: Funk & Wagnalls, 1983.

Gifford, Bill. "On the Borderline." *City Paper.* Vol. 13., no. 10, March 12–18. Washington, D.C., 1993. 24.

Goolrick, John T. *The Story of Stafford.* Stafford, VA: Cardinal Press, 1988.

Green, Constance McLaughlin. *Washington Village and Capital, 1800–1878.* Princeton: Princeton U P, 1962.

Hamlin, Talbot. *Benjamin Henry Latrobe.* New York: Oxford University Press, 1955.

Heaton, Ronald E., and James R. Case. *The Lodge at Fredericksburg, A Digest of the Early Records.* Bloomington: Pantagraph Press, 1975.

Heron, James Henry. *A Historical Sketch of Fredericksburg Lodge No. 4.* Fredericksburg: Kishpaugh Press, 1932.

Hickey, Donald R. *The War of 1812: A Forgotten Conflict.* Urbana: University of Illinois Press, 1989.

Bibliography

History of the White House. London: Odyssey Publishing & Mallard Press, 1991.

Historic American Buildings Survey, National Park Service, Department of the Interior. (District of Columbia Catalog). Complied by Nancy B. Schwartz. Charlottesville: University Press of Virginia, 1974.

Hunsberger, George S. "The Architectural Career of George Hadfield." *Records of the Columbia Historical Society.* Vol. 5152. Washington, D.C.: Columbia Historical Society, 1955. 46–65.

Israel, Fred L., ed. *The State of the Union Messages of the Presidents, 1790–1860.* Vol. 1. New York: Chelsea House Press, 1966.

Kapsch, Robert James. "The Labor History of the Construction and Reconstruction of the White House, 1791–1817." diss., University of Maryland, 1992.

Kite, Elizabeth S., *L'Enfant and Washington, 1791–1792.* Baltimore: The Johns Hopkins Press, 1929.

Latrobe, Benjamin Henry. *The Papers of Benjamin Henry Latrobe: Correspondence and Miscellaneous Papers.* eds.. John C. Van Horne and Lee W. Formwalt. Series 4, vol. 1. New Haven: Yale University Press for Maryland Historical Society, 1984.

———. *The Papers of Benjamin Henry Latrobe: Correspondence and Miscellaneous Papers, 1805–1810.* Ed. John C. Van Horne, series 4, vol. 2. New Haven: Yale University Press for Maryland Historical Society, 1986.

———. *The Papers of Benjamin Henry Latrobe: Correspondence and Miscellaneous Papers, 1811–1820.* Ed. John C. Van Horne, series 4, vol. 3. New Haven: Yale University Press for Maryland Historical Society, 1988.

———. *Latrobe's View of America, 1795–1820: Selections from the Watercolors and Sketches.* Eds. Edward C. Carter II, John C. Van Horne, and Charles E. Brownell. New Haven.: Yale University Press, 1985.

——— *The Journals of Benjamin Henry Latrobe: 1799–1820 Philadelphia to New Orleans.* eds. Edward C. Carter II, John C. Van Horne, and Lee W. Formwalt. New Haven: Yale University Press, 1980.

Lewis, David L. *District of Columbia, A Bicentennial History.* New York: W. W. Norton & Co., Inc., 1976.

Lipscomb, Andrew A., ed. *Writings of Thomas Jefferson.* Vol. 19. Washington, D.C.: Thomas Jefferson Memorial Association, 1903.

Madison, Dolley. *Memoirs and Letters of Dolly [sic] Madison.* Ed. Beverly Cutts, Port Washington, NY: Kennikat Press, 1971.

Madison, Dolly [sic]. *Memoirs and Letters.* Edited by her grandniece, Port Washington, New York: Kennikat Press, 1971.

Malone, Dumas. *Jefferson and the Rights of Man.* Boston: Little, Brown and Co., 1951.

McCue, George. *The Octagon.* Washington, D.C.: American Institute of Architects Foundation, 1976.

McCullough, David. *Truman.* New York: Simon & Schuster, 1992.

McDaniel, James I. "Stone Walls Preserved." *White House History Journal.* Vol. 1, no. 1. Washington: White House Historical Association, 1983.

Bibliography

McKee, Harley J. *Introduction to Early American Masonry: Stone, Brick, Mortar and Plaster.* Washington, D.C.: National Trust for Historic Preservation, The Preservation Press, 1980.

Morgan, J. D. "Major Robert Brent, First Mayor of Washington City." *Records of the Columbia Historical Society.* Vol. 2. June 1897. Washington, D.C.: Columbia Historical Society. 236–239.

Morris, Phillip. "Our Capitol-in-a-Meadow." *Southern Living.* Vol. 27, March 1992. 66.

Morris, S. Brent. *Cornerstones of Freedom; A Masonic Tradition.* Ed. John W. Boettjer. Washington, D.C.: The Supreme Council, 33o, S.J., 1993.

Myers, Gibbs. "Pioneers of the Federal Area." *Records of the Columbia Historical Society.* Vol. 44–45. Washington, D.C.: Columbia Historical Society, 1944. 127–59.

Nelson, Lee. *White House Stone Carving: Builders and Restorers.* Washington, D.C.: U.S. Government Printing Office, 1992.

Niemcewicz, Julian. *Under Their Vine and Fig Tree: The Travels through America 1797–1799.* 1805. Ed. Metche Budka. Elizabeth, NJ: Grassman Publishing Co., 1965.

Norton, Paul F. *Latrobe, Jefferson and the National Capitol.* New York: Garland Publishing, Inc., 1977.

O'Neil, Edward. *Repair and Maintenance of Masonry Structures: Case Histories.* Washington, D.C.: Army Corps of Engineers, 1995.

Osborn, John Ball. "The First President's Interest in Washington as Told by Himself." Records of the Columbia Historical Society. Vol. 4. Washington, D.C.: Columbia Historical Society, 1900. 173–198.

———. "The Removal of the Government to Washington." *Records of the Columbia Historical Society.* Vol. 3. Washington, D.C.: Columbia Historical Society, 1900. 136–179.

Owen, Robert Dale. *Hints on Public Architecture.* Washington, D.C.: 1849.

Padover, Saul K. ed. *Thomas Jefferson and the National Capital.* Washington, D.C.: Government Printing Office, 1946.

Pettijohn, F. J., Paul Edwin Potter, and Raymond Siever. *Sand and Sandstone.* New York: Sprinnger-Verlag, 1972.

The Register of Overwharton Parish, Stafford, Virginia 1723–1758. Compiled by George Harrison Sanford King. Easley, SC: Southern Historical Press, 1961.

Report on the Commission on the Renovation of the Executive Mansion. Compiled by Edwin Bateman Morris. Washington, D.C.: U.S. Government Printing Office, 1952.

Reps, John W. *Washington on View, The Nation's Capital Since 1790.* Chapel Hill: University of North Carolina Press, 1991.

Royall, Anne Newport. *Sketches of History, Life, and Manners in the United States.* 1826. New York: Johnson Reprint Corp., 1970.

Rutland, Robert A., ed. *The Papers of George Mason, 1725–1792.* Vol. 1. Chapel Hill: University of North Carolina Press, 1970.

Sale, Edith Tunis. *Interiors of Virginia Houses of Colonial Times.* Richmond: William Byrd Press, Inc., 1927.

Bibliography

Seale, William. *The President's House: A History*. 2 vols. Washington, D.C.: White House Historical Association with the cooperation of the National Geographic Society, 1986.

Smith, John. *The Complete Works of Captain John Smith, 1580–1631*. Ed. Philip L. Barbour, vol. 1. Chapel Hill: University of North Carolina Press, 1986.

Smith, Margaret Bayard. *The First Forty Years of Washington Society*. Ed. Gaillard Hunt. New York: Frederick Ungar Press, 1965.

Stafford Co., Virginia, 1800–1850. Census compiled by A. Maxim Coppage and James William Tackitt. Concord, CA: A. M. Coppage, J. W. Tackitt, 1980.

State of the Union Messages of the Presidents, 1790–1860. Ed. Fred L. Israel, vol. 1. New York: Chelsea House Press, 1966.

Steiner, Bruce E. "The Catholic Brents of Colonial Virginia." *Virginia Magazine of History and Biography*. LXX Oct., 1962. 399–403.

Stewart, John. "Early Maps and Surveyors of the City of Washington, D.C." *Records of the Columbia Historical Society*. Vol. 2. Washington, D.C.: Columbia Historical Society, 1895. 48–71.

Studebaker, Marvin F. "Freestone from Aquia." *Virginia Cavalcade*. IX, no.1, summer 1959. 35–41.

Thornton, Anna Maria Brodeau. "Diary of Mrs. William Thornton, 1800–1863." *Records of the Columbia Historical Society*. Vol. 10. Washington, D.C.: Columbia Historical Society, 1907. 88–226.

Tindall, William. *Standard History of the City of Washington from a Study of the Original Sources*. Knoxville: H. W. Crew & Co., 1914.

Truman, Harry S. *Off the Record; The Private Papers of Harry S. Truman*. Ed. Robert H. Ferrell. New York: Harper & Row, 1980.

Truman, Margaret, with Margaret Cousins. *Souvenir: Margaret Truman's Own Story*. New York: McGraw-Hill, 1956.

Washington, George. *The Diaries of George Washington*. Eds. Donald Jackson and Dorothy Twohig. Vol. 4. Charlottesville: University of Press Virginia, 1976–1979.

———. *George Washington's Ledger*. Washington, D.C.: Library of Congress.

———. *Writings of Washington*. Ed. John C. Fitzpatrick, vols. 32, 33, 34. Washington, D.C.: U.S. Government Printing Office, 1939–40.

———. "Writings of Washington Relating to National Capital," *Records of the Columbia Historical Society*. Vol. 17. Washington, D.C.: Columbia Historical Society, 1914. 3–232.

Waterman, Thomas Tileston. *Mansions of Virginia, 1706–1776*. Chapel Hill: University of North Carolina Press, 1945.

Whiffen, Marcus and Frederick Koeper. *American Architecture, 1607–1976*. Cambridge, MA.: The MIT Press, 1981.

Woodward, Fred E. "A Ramble along the Boundary Stones of the District of Columbia with a Camera." *Records of the Columbia Historical Society*. Vol. 10. Washington, D.C.: Columbia Historical Society, 1907. 63–87.

Index

A
Abraham, Spencer, 171
Accokeek Furnace, 23
Adams, Abigail, 7, 75-76
Adams, John Quincy, 77, 100, 109, 168
Adie, Hugh, 90
Alexandria, Virginia, 12, 14, 17-19, 41, 67, 177, 179, 189
Allen, William C., 41, 153
Alley, Shelton, 172
Anacostia River, 64-65, 144
Andrei, Giovanni, 97-99
Aquia Creek, 11-13, 26-27, 31, 56, 59, 61, 63, 70-71, 80, 86-90, 130-133, 151-153, 162-163, 175-176, 179, 185-187
Aquia Harbour, 10-12, 169
Aquia Quarries / public quarries, 34, 40-41, 50, 53, 70, 85, 93, 117, 161, 179
Aquia Overlook, 133
Aquia Stone. *See freestone*
Annapolis, 17
Architect of the Capitol, 31, 36, 61, 68, 90, 98-103, 108, 112, 117-119, 125-126, 128-130, 143-144, 153, 161, 165, 167, 170, 174, 187
Arnebeck, Bob, 64
ashlar, 70, 182
Association for the Preservation of Virginia, Antiquities (APVA), 152, 182
Atherton, Charles, 169
Atkins, Elmer, 156
Austin Ridge, 172
Austin Run, 59, 80

B
Baker, John W., 91
Baltimore, Maryland, 188
Banneker, Benjamin, 18-19, 39
Baptists, 51
Barton, Clara, 186
bas reliefs, 99-100
Battle of New Orleans, 109
Bigelow, Abijah, 156
Bill of Rights, 14, 86, 173, 177
bill stone, 53-54, 70
Black, Reed, 145
Blagden, George, 68, 70, 87-89, 115, 124-125, 127
Bladensburg, 105-106
Blair House, 136, 138
bluestone. *See foundation stone*

Boston, Massachusetts, 46, 115, 119
Botanical Gardens, 65
Boucher, Jack E., 61, 159, 178, 182-185, 187
boundary markers, 16, 18-20, 29
Braddock, General, 40, 177
Braddock's Rock, 40
Bramel, John, 91
Brent, Daniel Carroll, 14, 26, 85
Brent, George, 12, 26-27, 30, 49, 179
Brent, Giles, 12
Brent, Margaret, 12
Brent, Mary, 12
Brent, Robert, 26, 50, 83
Brent's Island. *See Government Island*
brick Capitol, 112
Brinkley, David, 165
British, 73, 105-107, 109, 113-114, 118
Brown, Robert, 115
Bryan, C. A., 147
Bulfinch, Charles, 102, 127
   dome. *See dome, Capitol*
Bush, President George H.W., 161, 163

C
Calhoun, John C., 125
canal, 27, 31, 41, 60-62, 65, 88, 129-130, 185
Cannon, Clarence, 136
Capellano, Antonio, 99
capitals, 46, 71, 88, 101, 113-114, 117, 127
   corncob, 98
   tobacco, 118, 127
carpenters, 47, 51, 71, 115, 119
Carroll, Daniel, 18, 26-27, 40, 86
Carroll, Mr., 106
Carter, President Jimmy, 157-58
Carter, John, 90
carvers, stone, 46-47, 71-72, 93, 98, 102-103
Carlyle House John, 177
cast iron, 126, 129-130, 132, 141, 152
Causici, Enrico, 99-100
Charles II, King of England, 12
Chesapeake and Ohio Canal (C&O), 130, 185
chimney pieces, 88, 179, 181
Christ Church, 14, 179, 181
Chopawamsic Creek, 70
Civil War, 62, 131-132, 147, 186
Clagget, Mr., 24
Coal Landing, 60
Cockburn, Rear Admiral, 106-107

Index

Coles, Edward, 105
columns, Capitol's 91, 123-127
Commissioners, the, 17-18, 45-48
Commissioner of Public Buildings, 124
Congress, 17-18, 39, 45, 61, 64, 98, 106, 109, 112, 120, 124, 126, 130, 132, 136, 144, 152, 156, 168, 170, 182, 184
Conner, Jane, 170
Conservation Fund, 169
Continental trench, 41
Conway, Thomas Barrett, 88
Cooke and Brent, 70-71, 80, 85-86
Cooke, George, 90
Cooke, John, 85-86, 90
Cornerstone Ceremonies, 38, 40, 43, 47
Cutts, Richard, 108

D

Daughters of the American Revolution (DAR), 20-21, 143
Davis, Congresswoman Jo Ann, 171-172
Dent's Landing, 90
derricks, 56, 94
Dickerson, Benjamin, 60
Dickerson, Milton, 59-60
District. *See Washington, D.C.*
Dome, Capitol, 108, 120-121, 126-127, 129-132, 143, 152
   Bulfinch dome, 120-121, 124, 127
drags, 59, 64
dressed stone, 53
Dumfries / Dumphries, 26, 30, 87, 181
Dunbar, John, 87-88, 91
Duron Paints, 157

E

Eastern Branch, 64-65, 88
Eck, Francis, 169
Edrington and Stone, 89-90
Edrington, John Catesby, 89-90
Eisenhower, Dwight D., 144
Elgar, Joseph, 60, 124
Ellicott, Andrew, 18-19, 29
England, 12-13, 36, 46, 70, 79, 106, 108
Ensign, William L., 170
Episcopalians, 51
Evelith, Isaac, 90

F

Fairfax County, Virginia, 177
Falmouth, Virginia, 132
Federal City. *See Washington, D.C.*
Federal Hall, New York, 173
Ferry Farm, 23
Feinburg, Dr., 157
Filmore, Millard, 168
Floyd, Governor John, 128
food, 47, 49-51, 69, 89, 133
Ford, Margaret Wall, 132
Fort Belvoir, 138, 181
Fortress Monroe, 185
Fort Myer, 138
Fossella, Vito, 173
foundation, 37, 39-42, 70, 80, 82, 87, 130, 136, 157, 165, 167, 172, 174, 182
foundation stone / bluestone, 39-41, 87
France, 47, 99
Franzoni, Carlo, 115
Franzoni, Giuseppe, 97-98
Fredericksburg, Virginia, 12, 23, 71, 86, 89, 174, 181, 188-190
Fredericksburg Area Museum and Cultural Center, 190
freestone / Aquia stone / sandstone, 9-11, 13-14, 23-24, 26-27, 31, 36-37, 40, 43, 46, 56, 59, 68-79, 85, 90, 93, 95, 99-100, 102, 114-115, 117-125, 129-131, 133, 137, 139, 144-145, 151, 157, 160-161, 166, 168, 174-177, 179, 184-186, 191
   definition, 9
   deterioration, 130, 151-153
   transportation, 17, 63, 65
   weight, 53, 95, 131, 152, 174

G

Garrett, Ethel Shields, 145
Georgetown, 17, 27, 34, 39-40, 43, 48, 54, 87, 185
George Washington Masonic National Memorial, 43
George Washington Stone Corporation, 133, 187, 191
Germany, 47
Gevelot, Nicholas, 99-100
Ghent, Belgium, 109, 183
Gibson, John, 26, 30, 61, 85
Goldsworthy, Andy, 173
Goolrick, John T., 133

Index

Goose Creek, 63
Gordon, George, 62
Government Island / Brent's Island / Wiggington's Island /Island quarries, 10-11, 49, 54-56, 61, 63, 72, 80, 85, 88, 90, 129, 132, 147-148, 165, 169, 171, 173-175
Greenleaf, John, 69
Griffis, John, 90
Gunston Hall, 177-178, 189

H
Hadfield, George, 70, 73, 111-112, 115
Hall, Jim, 133
Harris, Gayle, 21
Henry, John & Company, 87
Historic American Buildings Survey (HABS), 115, 159
Hoatling, Ed, 48, 170
Hoban, James, 34-37, 40-41, 48, 51, 68-69, 71, 73, 115, 119, 155
Hodgkins, George, 165
Holland, 47
holidays, 50
Holliday, Frederick W. M., 147
Hore, John, 91
Horton, Cossom, 90
Horton's Landing, 90
Horton, Lucy, 90
House Chamber, 106
House of Representatives, 51, 77, 98, 108, 119, 168, 172-173
Howell, Samuel B., 147

I
Irish, 35, 48, 50, 69, 71
Island quarries. *See Government Island*

J
Jackson, Andrew, 109, 186
Jefferson Obelisk, 36
Jefferson, Thomas, 9, 25, 29, 36, 56, 79, 81, 97-98, 109, 113
Jeffreys, Herbert, 11
Johnson, Joseph, 90
Johnson, Thomas, 18, 40
Jones Point, 18, 20
Joslyn Art Museum, 190
journeyman, 47

K
Kapsch, Dr. Robert, 103, 115, 159, 170

Keener, Marcia, 170
Kenmore, 174, 181-182, 184, 189
Kennedy, John F., 144
Key of Keys, 40, 87
Kimball, Fiske, 179

L
laborers. *See workers*
Lafayette, Marquis de, 168
Lamb, Mary, 91
Lamb, William, 91
Lambert, William, 36
Lancaster, 15
Lane, Samuel, 116, 119
Latrobe, Benjamin Henry, 31, 41, 61, 65, 79, 83, 86, 88-90, 97-98, 101, 112-115, 117-119, 124, 127, 133, 171
Lawyers Title Insurance Company, 147-148
Lee, Robert E., 179
Leinster House, 36
L'Enfant, Peter (Pierre) Charles, 9, 23, 25-27, 29, 31, 34, 39-40, 49, 61, 65, 85, 88, 175-176
Lennox, Peter, 119
Lewis, Betty, 181
Lewis, Fielding, 174, 181
Lewis, John, 170
Library of Congress, 45, 77, 130
limestone, 95, 136
    Indiana, 167, 177, 191
Lincoln, Abraham, 186
Lincoln, Blanche, 171
Lincoln Memorial, 40
liquor, 50-51, 73, 155
Loker, Katherine B., 173
Loth, Calder, 170

M
Madison, Dolley, 105-106
Madison, James, 27, 105, 108, 168, 183
Mallow, Wendy, 170
marble, 43, 56, 65, 98-100, 103, 114-115, 117, 125, 129-130, 136, 143-144, 153, 166, 168, 190
    Georgia, 144
    Italian, 114, 117
    Tennessee, 130
Marine Band, 31
Marshall, Chief Justice John, 184
Maryland, 12-13, 17-20, 42, 47, 88, 91, 114, 130, 188-189, 201, 204, 206, 208, 212, 216

Index

Maryland House of Delegates, 88
Mason, George IV, 14, 86, 173, 177, 181
Mason, Mary Thomson, 86
Masonic lodges, 42-43, 140
masons, 31, 41-42, 46-47, 49-51, 53, 69-70, 77, 93-94, 96, 117, 161, 163
    ceremonies, 40, 43, 47
    freemasons, 18, 42, 46
    masons' marks, 140-141
    master masons, 46
    rough, 46, 93
Massachusetts Avenue, 69
McDaniel, James I., 160
McKee, Harley, 56
Meridian Pier, 36
Methodists, 51
Metts, Thomas, 148
Millikin, 56
Millar's Quarry, 89
Moncure, John, Jr., 91
Moncure, William A., 89
Monroe, James, 116, 168
Monticello, 33, 98
Monumental Church, 184
mortar, 41, 64, 71, 95, 130-131, 152, 165
Morton, James, 54
Mount Airy, 108, 178-179, 183
Mount Vernon, 15, 23-24, 42, 86, 138, 165, 183-184
Mounty, John, 54
Muir, James, 18
Myrtle Grove, 90

N
National Arboretum, 145
National Archives, 45, 186
National Cathedral, 160-161, 191
National Portrait Gallery and the National Museum of American Art, 186
National Park Service, 153, 159-160, 169-170
National Register of Historic Places, 173
Nelson, Lee, 54
Nelson, Thomas House, 175
New York City, 17, 25, 173, 188
Niemcewicz, Julian, 48-49, 73
Nixon, Richard Library and Birthplace, 173

O
Octagon, The (Museum), 108-109, 179, 182-184
Oehrlein and Associates, 145
Old Cape Henry Lighthouse, 152
O'Neale, William / O'Neil, 64, 87
Oven, The, 77
Owen, Robert Dale, 129
oxen, 39, 62-64, 80, 126

P
Page, Russell, 145
Palumbo, Vincent, 160
Patent Building, Old, 89
Peale, Charles Wilson, 80, 97
Peale, Rembrandt, 34, 100
Pearce, William, 87
Pearson, Drew, 65-66
Perley and Edrington, 89-90
Peyton and Cooke, 90
Peyton and Dent, 90
Peyton, Rouzee, 91
Philadelphia, Pennsylvania, 14, 17, 30-31, 34, 39, 46, 56, 72, 81-82, 102, 109, 184, 188
Plunkett, Patrick, 96, 161-163
Pohick Church, 14, 179-181
Poplar Point Nursery, 144
Porter, Charles, 89
Potomac Bluestone, 40
Potomac Lodge, 43
Potomac River, 9, 11, 17, 40, 42, 61, 65, 177
Presbyterians, 51
President's House. *See White House*
Prime Meridian, 36-37
Princeton, 17
public quarries. *See Aquia Quarries*

Q
quarries, 10-12, 24, 26-27, 30-31, 34, 40, 46, 48, 5_, 51, 53, 56, 59, 63, 67-68, 70, 79-80, 85-86, 88-8_, 93, 114, 117, 123-124, 127-130, 133, 161, 175, 179, 185-186
quarry clothes, 51
quarrying, 27, 54, 56, 61, 79, 124, 128, 165
quartz, 18, 80
quicklime, 95
quoins, 13, 131, 176-177, 179, 181

R
Rabaut, Louis C., 137
railroad, 61-62, 133
RAMCO, 157, 163
Rappahannock River, 11, 23, 89, 132, 181
Rappahannock sandstone, 182
Rayburn, Sam, 144
Reagan, President Ronald, 163, 166

religions, 13
repair cost / rebuilding, 111
Residence Act, 45
Revolutionary War, 17-18, 175, 181
Roberdeau, Issac, 30-31
Roberts, Jonathan, 109
Robertson, A. Willis, 166
Robertson—Towson House, 172
Robertson, William, 63, 80-81, 88, 133
Rock Creek, 40, 63
Rock Rimmon, Rock Raymond, 88-89
Roe, Cornelius Mc Dermott, 24, 69-70, 116
Roman Catholic, 13, 18
Rotunda, Capitol, 77, 99-100, 102, 118, 121, 128, 131, 168
Royall, Anne Newport, 48, 121, 126
rubble stone, 40-41, 53
Rudder, Thomas, 163, 167
RUDCO, 157-158
Rush, U.S. Attorney, 107
Rush, William, 97

S

scabble stone, 53
scaffolding, 51, 73, 94, 96, 100, 111, 157, 163
schooner, 89
Scouten, Rex, 140, 153, 156, 159, 167, 169, 172
scows, 31, 41, 59-60, 69-70
Scotland, 47, 96
Seale, William, 51, 170
Segar, Wilbur, 56
Seneca stone, 114, 117, 130, 185
Senate, 77, 98, 109, 111, 113, 143, 166, 168, 171
Senate chamber, 113, 119, 120
Serurier, Louis, 106
ships, 63, 133, 175, 186
Shriber, Aaron, 173
Simons, Farrar A., 148
skids, 69, 126
slacking, 39, 95-96
slaves / slavery, 45, 47-49, 51, 64, 67, 69-70, 116, 128, 133, 170, 174
sledges, 59
sloop, 63
Smith, Captain John, 9
Smith, James, 63
Smith, Margaret Bayard, 108-109, 112-113, 168
Smith, Samuel Harrison, 168
Smith, Congressman William, 63

Smithsonian Institution, 129-130, 152
Statue of Freedom, 132
Stone, Elizabeth Hawkins, 90
Stone, Governor William, 91
Society of Friends, 51
Stafford / Stafford County, 9, 13, 15, 23, 27, 59, 62, 80, 86, 89-91, 124, 127-128, 132-133, 148, 169, 174, 176, 179
Stafford County Board of Supervisors, 169
Stafford County Courthouse, 13
Steuart (Stewart, Stuart), Robert, 26, 59, 72, 88, 148, 173
Steuart's Wharf, 59-60, 62-63
Steuart, William, 88, 126, 172
Stewart, J. George, 143-144, 165-166
stone boats, 59
Stone, Caleb, 24
stone carvers, 46-47, 71-72, 93, 98, 102-103
stone cutters, 31, 41, 46-48, 50, 72-73, 103, 114-116, 125, 127
Stone, James W., 90-91
stone setters, 46
Stuart, David, 18
Supreme Court / old chamber, 77, 98, 112, 168
Suter's Tavern, 27
Suttle, Agnes, 90
Suttle, John H., 90

T

Tallifur, F. W., 91
Tayloe, Mrs. Henry Gwynne, 108, 179
Tayloe, John III, 183
temple, 47, 77
Thornton, Mrs. William, 76, 96, 107
Thornton, William, 36-37, 70, 76, 108-109, 121, 123-124, 183-184
Tiber Creek, 63
tombstones, 13, 91, 133, 184
tools, 46-47, 54-56, 89, 162-163
Towson House, 172
Towson, Thomas, 60, 90-91, 124-125, 172
Towson, Thomas P., Jr., 132
Treasury Building, 51
Trenton, 17
Trig, Clemet, 24
Truman, Harry, 135-137, 139-141, 143
Truman, Margaret, 135, 140
Trumbull, John, 70
Turner, Nat, Rebellion, 128
Tyler, William, 90

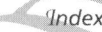

Index

U

U.S. Capitol, 9, 23-24, 29-33, 36-37, 41-43, 45, 47-50, 56, 60-61, 64-65, 68-73, 76-77, 79-83, 88-91, 94, 97-103, 107-109, 111-120, 123-127, 129-130, 133, 140, 143, 145, 151-153, 157, 161-162, 165-168, 170-175, 183, 187
    East Front extension, 60, 91, 103, 123-124, 126-127, 143-145, 152, 161, 173
    West Front restoration, 124, 127, 166, 172
United States Supreme Court Building, 112

V

Valaperta, Giuseppe, 115
Virginia Declaration of Rights, 14, 86, 173, 177
Virginia House of Delegates, 86
Virginia Landmarks Register, 173

W

Wakefield, 21
Walker, George, 17, 27
Waller and Morton, 124
Waller, Withers, 90, 124
Walter, Thomas U., 130
Walters, Gary J., 161
War of 1812, 105, 151, 168, 183
Washington, D.C. / Federal City / District, 9, 13, 15, 19, 27, 36, 47, 56, 63, 65, 83, 105, 129, 147, 157, 165, 170, 183-184, 186, 188-189, 191
Washington, George, 9, 14, 17, 23, 30, 33, 36, 40, 43
Washington Monument, 36, 172, 184
Watts, J. C., 170
Westmoreland County, Virginia, 23
Whelan, Patrick, 41
White, George M., 161
White House / Presidential Mansion, 7, 9, 21, 24, 33-36, 39, 40-42, 43, 45, 48-51, 53, 56, 63, 64, 68-73, 75-76, 77, 80, 83, 88, 93-94, 96, 101-103, 105-108, 111-112, 115, 124, 140, 155, 168
    renovation, 135-137, 140-141, 160, 167
    painting, 73, 145, 155-157, 159, 160
    stone restoration, 160-161, 163
White House Historical Association, 171
Whitman, Walt, 186
Widewater, 90-91
Wiggington, William, 12-13
Wiggington's Island. *See Government Island*
Williamson, Collen, 39-41, 51, 67-69, 71, 102, 117
Williams, C. M., Jr., 169-170
Williams, Mr., stone provider, 41

Winslow, Lorenzo, 141
Wolff, Rick, 172
Woodlawn Plantation, 183-184
Woodstock Plantation, 12-13, 132, 179, 189
Wood, Sweet, and Swofford, 152
workers / laborers 39, 45-51, 56, 68-69, 72-73, 79, 82, 94, 116, 119, 127, 136, 162, 170
    apprenticeship, 47, 69, 116
    food, 47, 49-51, 69, 89
    hours, 116, 119, 166
    housing, 49
    laborers, common, 45, 47, 116
    laborers, unskilled, 47
    wages, 45, 49-50, 67-69, 71-73, 83, 86, 97, 116, 170
Wren, James, 179, 181
Wright, William, 67

Y

York, 105